J. W Bright

Cancer

Its Classification and Remedies

J. W Bright

Cancer
Its Classification and Remedies

ISBN/EAN: 9783337125394

Printed in Europe, USA, Canada, Australia, Japan

Cover: Foto ©berggeist007 / pixelio.de

More available books at **www.hansebooks.com**

CLASSIFICATION AND REMEDIES.

BY J. W. BRIGHT, M. D.

PHILADELPHIA:
PUBLISHED BY S. W. BUTLER, M. D.
1871.

Entered according to Act of Congress in the year 1871,

BY S. W. BUTLER, M. D.,

In the Office of the Librarian of Congress at Washington.

INTRODUCTION.

In offering the following pages to the profession, my aim has been to place in their hands a work that will save them much labor and toil in collecting facts, and at the same time give a Classification that will remove the difficulty of having so many names for the same form of Cancer. If these efforts shall have the effect to stimulate others of the profession to a further effort in the development of truth, and so advance the science of the healing art in the treatment of this dire disease, I shall be fully compensated for my labor.

THE AUTHOR.

Lexington, Kentucky, July, 1871.

CONTENTS.

Introduction,	7
Preliminary Observations,	9
Treatment of Cancer by the Most Eminent Physicians and Surgeons,	66
Tubercle,	85
Cancer Defined,	94
Classification of Cancer,	107
Preparation and Composition of Remedies, and Record of Cases,	118
Views of Various Authors,	140
Sprouting Cauliflower Cancer,	150
Landolfi's Method,	154
Cancer on Internal Organs,	170
Note on Cundurango,	183
Index,	189

CANCER:

ITS CLASSIFICATION AND REMEDIES.

Almost simultaneously with my entrance into the medical profession, I was called upon to treat Cancer. I knew nothing of its treatment except what I had learned from the books, and that was meagre. Those who had written on cancer had but a limited knowledge of that disease. Various remedies however were prescribed, but the knife was generally relied upon, and almost universally used. Adopting the most popular mode of treatment, I used the knife in every case where I could apply it: but where I thought it advisable, I tried the usual remedies I found prescribed, such as arsenic, cicuta, and in a few cases the actual cautery. I also used such constitutional remedies as my judgment suggested, as a purifier of the blood, and for the support of my patient's strength; for it must be remembered, at that day (fifty-six years ago,) cancer was supposed to be a blood disease, and in every case primarily constitutional. And even now many physicians and surgeons, otherwise eminent in their profession, still hold to the old theory, that cancer is a blood disease. But the light of science, developed by a more perfect knowledge of physiology, has brought to light truths that were not known to our ancestors. A more perfect knowledge of the tissues, their functions, the circulation of the fluids, and so forth ; the formation of cells, and of tubercles, their location, functions, persistence, or non-persistence, their character as it pertains to malignancy, or non-malignancy, has, by repeated search and research, been so clearly and fully demonstrated, that we are now, not left to grope our way in the dark, or rest our opinions and practice upon uncertain conjecture, but have the light of demonstrative truth laid open to our view, leaving no doubt as to the real nature of cancer.

These facts demonstrate to our view the varieties of Cancer, and lead us to their proper classification. When these questions once become settled and established, the mind will be relieved from the embarrassment of so many different names for the same variety of this disease, and which have usually been applied to it from the particular location in which it is found, or the appearance of the matrix in which the cells are formed. When we are freed from these difficulties, and the classification is simplified, and each variety clearly defined, many of those barriers will be removed that have retarded our progress in deciding upon the proper mode of treatment to be pursued in this disease, which must, however, necessarily vary according to the classification of the disease, and its location. If the disease, for instance, is located in the tongue, although it may be epithelioma, you cannot use the same remedies that you can on an outer surface, and if it be a fungoid, the remedy must be different from that used on the lip, or on the mamma. I never saw a hematoid cancer in the tongue, nor do I believe that this variety of cancer can appear there. We shall, however, have more to say on this subject, when we come to describe the several varieties, and give the appropriate mode of treatment for each.

As I stated in the outset of these remarks, my first treatment of Cancer was mainly with the knife, with what success you shall presently hear. My first cases were located in the cheek and lips. I removed them, as I thought, by excision, cutting out the sore and a small portion of the margin embracing the hard and everted edges, and going as deep as I thought was necessary to clear the base of all cancerous matter. A majority of the cases would heal kindly, and remain for a time to all appearance cured, and the cancer eradicated. To my mortification, however, in from six months to two years, a small pustule, or a little dry sore would appear in the cicatrix, which would spread and destroy the contiguous parts more rapidly than the former ulcer had done. Most frequently these little pustules would appear on the margin of the sore. I resorted to the knife the second and even the third time, but to my extreme annoyance, all my efforts proved abortive in nine cases in ten: and my patients succumbed to the disease. In some rare instances where the knife was used in the early stage of the disease, and the whole of the diseased parts were removed, the operation proved a success. I pursued this practice for about twenty years, but finally became dissatisfied and disheartened, gave up the knife, and turned my attention to a more thorough research into the nature of cancer, and a different mode of treatment.

In proceeding to give the various remedies which I have used in the treatment of Cancer, and their success, I shall notice the different theories offered by writers on this subject. In relation to the application of remedies from the vegetable kingdom—STORK eulogized the Hemlock as a cure for Cancer. Dr. OSBORN, of Dublin, in 1840, used this remedy, and he says, "even the extract, imperfect as it is, has an effect in appeasing the pain in Cancer of the Uterus:—and that without exerting sensible narcotism." This almost excuses STORK for the error into which he fell, in proclaiming it as a cure for cancer. Dr. OSBORN tried it externally and internally, with only mitigating effects, but no radical cure has been performed in his hands by it: though sometimes with remarkable alleviation of pain, even after opium had failed. He never observed any disastrous effects from it, except in one case in a lady who labored under Scirrhus of the Uterus. To obtain this end, however, he increased the dose to four grains three times a day. This produced headache and black motes floating before the eyes, and double vision, which all disappeared when the remedy was discontinued. The extract of Hemlock is an unreliable remedy, probably owing to the manner in which it is prepared.

The coneine which in the process of decomposition is partly resolved into ammonia, is in the ordinary preparations which have come under the observation of Professor GEIGER, more or less deficient and often entirely absent. As a test for this, it is only necessary to add some water of caustic potash, when the ammonia may be detected by the odor, or by holding over it a rod dipped in muriatic acid, whereupon the fumes formed by the muriate of ammonia become visible. This experiment will detect ammonia in all the impure extracts.

For those who may wish to use Dr. OSBORN's method of preparing a pure article, it is as follows:—Take the leaves and smaller branches of the stalks fresh gathered before flowering, pound and intimately mix them with an equal weight of treacle. This will keep for any reasonable length of time without decomposition. Prepared in this way, it is not narcotic, but an active purgative without pain or griping. But the only reliable preparation of hemlock is that prepared from the seed. For external use the electuary prepared as above will answer very well. But for internal use, the seeds should be decidedly preferred, for the natural process of ripening and drying preserves them from change or decomposition, consequently all their virtue is retained, and further, because the seeds of hemlock contain most of the active virtues of the plant.

M. LEFRANC, of Paris, wrote in 1843 thus: This excellent surgeon has noticed among the circumstances relating to Cancer, that it seldom attacks at once all the various tissues of the organ. Thus in the stomach, e. g., its ravages are sometimes confined to the mucous, or to the cellular, or to the muscular tunic of that organ; and even when all the tissues seem equally involved, a careful dissection will usually point out which had become first affected. So, too, in the examination of the bodies of a great number of women who died of cancer of the breast in the Salpétriére; the pleura, though in contact with the disease, was found unaffected. These and other facts induced him to inquire whether the ravages of cancer in an organ might not be curtailed, especially if its tissues formed part of any peculiar structure fitted to isolate the malady; if so, the removal of the superfices performed in time, might save the remainder of the organ. Let this suggestion not be forgotten.

LEFRANC first put this idea into practice in a case of Cancer of the penis. Here we shall see the result of the important thought. He began his operation with what he calls an exploratory incision, in order to examine whether the corpus cavernosum yet remained in a healthy condition, so as to admit of the organ being saved. Finding this to be the case, he removed the cancerous part by a long, painful and difficult operation, and the patient perfectly recovered, performing afterwards his generative functions in a satisfactory manner. In the next case, the testis and cord were laid bare by a laborious dissection; the penis almost detached from the body of the pubis, and the cancer pursued to the roots of the corpus cavernosum, which was exposed, and, as if prepared for an anatomical demonstration, here and there were evidences of cancerous deposition that had to be removed by the forceps and scissors; the surface afterwards being well scraped with a bistoury. The cure was perfect, and the extent of cicatrization such as can only be imagined by those unaccustomed to the wonderful reparative powers of nature.. Lefranc has treated many similar cases with like effect. He treats cancer of the tongue, vagina and rectum, after this manner. Cancer of the nose almost always spares the cartilages. He affirms, that the cicatrices are not unsightly, and by suppressing exuberant granulations with the nitrate of silver, very little deformity will ensue. This great surgeon is entirely convinced that cancer is not contagious. He also denies that there is any general infection of the system in this disease. This is true to a certain degree in the early stages of the disease, setting aside the old theory that cancer ab initio is a blood disease. He supposes the disease arises from a greater or less number of tubercles. He supposes these, like ordinary tubercles, may be more or less confined to certain organs. This is true, but those organs or tissues he has not pointed out; as suppuration attacks and removes tubercles, not simultaneously, but in successive attacks, so cancer might be removed by several successive operations.

Lefranc believed that cancer might be produced solely by common irritating agents. Thus, ulcers, cicatrices, accidental tissues, frequently become cancerous when irritated, while instances of relapse after the removal of such are very rare. Many persons are liable to pimples on the face consisting of accidental tissue; no change usually occurs in these until old age approaches, when they either spontaneously ulcerate, or do so in consequence of the irritation by external causes. The pus is scanty in quantity and concreting into a thick crust; the person takes no notice of it, but sooner or later the cancerous condition supervenes, if it has not already commenced with the first ulceration. Persons who are liable to the production of permanent pimples upon the nose, should be careful not to excite their cancerous degeneration by the two rough use of the pocket handkerchief.

Razor wounds in such subjects often produce dangerous ulceration. As soon as ulceration occurs, if it be not yet cancerous, he applies the nitrate of silver, and follows its use by mild stimulating dressings. But he has seen such frightful aggravations follow the caustic, that he advises the immediate use of the bistoury or the part to be touched with the protonitrate of mercury: with the intention of changing its mode of action, rather than eradicating the disease. Still, he says, in this way you may effect a cure, if you enjoin a strict antiphlogistic diet. Tissues of this type, when neglected, frequently degenerate into cancer: rarely do so if they are touched with the nitrate of silver. Those slight erosions of the cheeks, covered by thin whitish scales, are the first evidences of Epithelial Cancer, and may be readily and certainly removed by the timely application of the proper remedies. M. LEFRANC mentions that all experienced surgeons must deny that cancer is always well characterized. In this way he explains DESAULT's supposed cures of cancer of the rectum, and cites cases of apparently incurable cancer yielding to means seemingly inefficient. He observes also that we pronounce many ill-looking sores to be cancerous merely from their localities, and that we should never so consider them if seen on other parts. It is well known that syphilis often simulates cancer, antiphlogistic remedies curing the sores, thus proving that they are not cancer.

LEFRANC places more reliance on an operation for the removal of diseased portions than most surgeons. He operates occasionally in the advanced stage; and even when the neighboring lymphatic glands are enlarged, he will not always allow them to be an impediment to the operation, trusting to their dispersion by antiphlogistic treatment, as all enlarged glands are not cancerous, or even scirrhous. In case of cancer, he says, " We are to consider the whole substance of the tumor as being occupied by scirrhous." While LEFRANC was pursuing his experiments in the cure of cancer by local bleeding, he observed that the tumors so situated underwent a notable diminution at their circumference. This induced him to make careful pathological examinations in some fatal cases; and although in a few of those the whole tumor seemed to be completely occupied by cancer or scirrhous, yet, in the great majority the morbid change was found in the centre. The scirrhous nature of the swelling gradually changed into the common inflammatory induration as the circumference was approached. From these facts he deduces some practical proceedings; thus, a cancer may be of too great an extent to admit of an operation, for the patient would succumb during the healing of the large wound necessary for its removal. Leeches in proportion to the strength of the patient must be applied around the base of the tumor, following them by a large cataplasm and enjoining a careful diet. As long as the constancy of the pain and the augmented heat denote the continuance of sub-inflammation these means must be repeated. The chronic nature of the disease and the enfeebled state of the patient, however, admonish us that the repetition must be neither too hasty nor rigorous. In a few days we may have recourse to resolvent applications. By such means carefully adapted and extended over a sufficient space of time, cancerous tumors have been frequently reduced at La Pitié, from a size represented by 10 to one of 6 to 4, and not only have the operations then been practicable, but, no more than an ordinary wound has remained after its performance.

Again, a Cancer may not be excessively voluminous, but it may involve such important parts as to render its removal difficult, or impossible, by the means mentioned. The lessening tumor returns from the vicinity of the important blood-vessels and nerves, which may have presented the obstacle to an operation. He says, some practitioners, observing this particular diminution of the tumor produced by these means, and ignorant of the principles upon which they have been recommended, reject the operation altogether, vainly believing they can dissipate the cancer altogether in the same way : thus giving time for the disease to make fatal progress. He says after the operation for cancer, the fingers must be carefully passed over the whole surface of the wound, by which means small tubercles easy of removal may be often felt: they sometimes being completely imbedded in the pectoral muscles. When very numerous, they can very easily be completely extirpated, and when left they give rise to the reproduction of the disease. In patients dying a few days after the operation, he has found hundreds of these bodies, even under the clavicle, scapula, and upon the pleura ; observation proves that cancer is liable to relapse in proportion as its progress has been active, or it has been complicated with active inflammation. This last is to be met by local depletion, continued some days prior to the operation. Relapses will be less likely to occur, also, if half an inch of sound skin be included in the incision, leaving however, enough integument when it is sound to procure union by the first intention, thereby diminishing the extent of irritation of the wound and subsequent cicatrization. The surgeon's care of his patient, however, must not terminate with the production of a cicatrization, but must be directed to the prevention of a relapse, which when patients are docile enough to obey the directions given, may usually be accomplished.

All causes, whether physical or moral, must be sought for and removed. In persons of a sanguine temperament, or those suffering from the suppression of some flux, let bleeding be practiced ; and if the parts in the vicinity of the cicatrix become congested, a revulsive bleeding of three or four ounces must be employed. The vitality of susceptible organs must be modified by the use of powdered hemlock, for many months. Beginning with a grain every morning, the dose may be gradually augmented to four grains. LEFRANC has a very high opinion of this remedy as a solvent and nervine. Its utility as a nervine is also well seen in gastralgia in women suffering from uterine affections. Occasionally mild aperients are needed in cases of retrocession, and after remitting the hemlock, the external or internal use of iodine. Much good also results from compression employed by means of a suitable compress, and a bandage extending beyond the cicatrix. The cicatrix must not be irritated, especially by too early movement of the parts. The mildest diet, and if debility be not present, even abstemiousness should be enjoined. We have thus given a synopsis of LE-FRANC's theory and treatment of cancer. The practice has been much improved in many regards since those days; and the theory of cancer presented in a very different light.

Dr. MONTGOMERY, physician to the King and Queen's Hospital, Ireland, and Professor of Midwifery, is of opinion that there is a state of Cancer of the Womb, which is remediable, and the germs which may be destroyed, a stage which precedes the two which are usually described by authors. The reason why this stage is not more generally recognized is, that the accompanying symptoms are so slight as to attract very little attention from the patient, and thus are suffered to remain without treatment, till a profuse hemorrhage or some violent fit of pain sounds the alarm. Then on examination the disease is found to have passed into the second stage. The surrounding tissues are indurated and consolidated with the organs concerned. There is no doubt but many of these cases are overlooked by practitioners, simply from not properly examining the vagina. Upon examination, the margin of the os uteri will be found hard and often slightly fissured, and it projects more than usual into the vagina, and is irregular in its form. In the situation of the muciperous glands, there are felt several small hard lumps or distinctly defined protuberances, almost like grains of shot or gravel lying under the mucous membrane. Pressure on these with the end of the finger gives pain, and the patient often complains that it makes her feel sick at the stomach.

The cervix in most instances is slightly enlarged, and harder than it ought to be. The circumference of the os uteri, especially between the projecting granules, feels turgid, and to the eye presents a deep crimson color, while the projecting points have sometimes a bluish hue. There is no thickening, or other alteration of structure in any part of the vagina, at its connection with which the cervix uteri moves freely. Nor is there any consolidation of the uterus with the neighboring contents of the pelvis. In fact, the morbid organic change appears to be at first entirely confined to the os uteri and lower part of the cervix. This stage of the disease is in many instances very slow, lasting sometimes for years before the second and hopeless stage is established. During this time the patient experiences only slight and transient attacks of pain, or perhaps only sensations of uneasiness, referred often to the situation of one or the other of the ovaries, or about the os uteri with an anomalous tingling along the outer part of the thighs, or on their inside. These last for a few hours, sometimes for a day or two, and then disappear, perhaps for weeks, but again and again return, in the same situation, and for a long time are not increased in severity. Sexual intercourse is painful, which she ascribes to some deep-seated part being touched, and the act is followed by the appearance of blood. There is frequently slight irritability of the bladder, but the appetite, digestion and sleep may for a long time continue good. The pulse generally gives no indication of the existing disease, or its changes; an observation which will be found applicable to many uterine affections of a graver character. In short, the general health may long remain quite undisturbed. Nor has the patient in many instances the slightest suspicion that there is any thing wrong with her, nor does she think of seeking medical aid till she is induced to do so by the solicitation of her husband, or some anxious friend who has become, as she thinks, unreasonably alarmed about her condition.

Dr. MONTGOMERY gives his pathological views of this disease as follows: He says, continued observation sufficiently satisfied me that in a great majority of instances the first discoverable morbid change which is the forerunner of cancerous affection of the uterus, takes place in and about the muciperous glands, or vesicles, sometimes called ova Nabothi, which exist in such numbers in the cervix and margin of the os uteri; these become indurated by the deposition of scirrhous matter around them, and by the thickening of their coats; in consequence of which they feel at first almost like grains of shot, or gravel, under the mucous membrane. Afterwards, when they have acquired volume, by further increase of morbid action, they give to the part an unequal, lumpy, or knobbed condition, like the end of one's fingers drawn closely together. When this second stage is established, all means hitherto devised have failed in producing any beneficial results. He diagnoses this disease as follows. He says, the only affection of the uterus for which this disease could be mistaken, and this only by carelessness, is the irritable uterus, from which, however, it is essentially different, inasmuch as it is accompanied by, and tends to produce still further changes in the structure of the organ, which, although unduly sensitive on examination, is not the seat of the exquisite tenderness and pain observed in irritable uterus; from which it also differs in having the increase of volume of the parts affected, well marked and constant until removed by treatment; and in the existence of the other organic alterations already enumerated, as well as in the different results of this affection.

The second or fully formed stage of Cancer of the Womb any one accustomed to examine the organ, will at once recognize. He says, in almost every instance the treatment should be begun by the local abstraction of blood, either by cupping or leeches applied directly to the os uteri, or as near as possible to that organ; and their application will in most cases require to be repeated, and should be accompanied in most cases by free use of anodyne fomentations.

With regard to venesection, although it may be desirable to practice it under particular circumstances, it is not in general required. And he says the case in which it is called for, should be regarded as an exception in the most suitable plan of treatment, except there be something specially forbidding its use. He gives mercury in small doses, so as to bring the system very gently, but decidedly under its influence: for which purpose he would combine it with iodine, in very minute proportions, with camphor, opium, hyoscyamus, or hemlock. Or, it may be introduced by friction, especially where there exists evidence of inflammatory action. Afterwards iodine, or iodide of potassium may be used, both externally and internally.

Iron will be found a beneficial and powerful agent, especially in the form of the saccharine carbonate, given in the nascent state. Arsenic has received the testimony of many able physicians in its favor, as an agent capable of giving great relief in this disease. MONTGOMERY also adds his testimony to the same effect, having obtained marked benefit from its use, especially when combined with anodynes, even in the advanced stage of this disease.

He says of the iodide of arsenic, he cannot speak from his own experience, but thinks from the nature of the compound, and still more from the success which appears to have attended its administration in cancerous affections, by Dr. A. T. THOMPSON and Dr. CRANE, of Canterbury, that we are justified in expecting that it will prove a useful remedy in this disease. Counter-irritation is an agent of great influence, and may be established in various ways,—which are unnecessary to enumerate,—but a very effectual mode is to establish a small blister over different parts in succession, and so keep up a free discharge for several days. The warm hip-bath is a means of great value throughout the entire treatment; the effect is soothing. The uterine irritation may be much ameliorated by admitting the warm water to come in contact with the internal surface of the vagina and os uteri, which may be accomplished without difficulty, by introducing into the vagina a suitable speculum made of wire-gauze, coated over with caoutchouc; or, a small plain metal speculum, with perforations in the sides, will answer the purpose very well. The patient can use this instrument herself, better than any one else could. I may observe, where warm baths are used in the treatment of amenorrhœa, this mode of managing them may be adopted with great advantage.

After the removal of congestion and organic changes from the os uteri, there remains occasionally a sensitiveness of the part which causes the patient much discomfort, and will be best relieved by the use of the bath, as above directed, conjoined with anodyne applications to the part, or the nitrate of silver in solution. He applies this by means of a bent glass tube, of about an inch in diameter, which the patient can introduce and manage herself. She should lie on her back and introduce the tube as far as its curvature, then pour into the upper end the medicated solution, which will immediately pass to the os uteri, and can be retained there as long as necessary, which is a great advantage over all other methods for its use. Every thing that would stimulate the uterus must be strictly prohibited: such as riding on horseback, dancing, etc., and especially sexual intercourse. The quality and quantity of the patient's diet should be strictly attended to. In all cases it should be light, but nourishing. Wine, if used at all, should be of a very mild kind, and very sparingly taken. All strong drinks should be strictly avoided. No circumstance connected with the treatment of this disease requires more scrupulous attention than the regulation of the patient's habits and mode of living. If this be not very carefully managed, all other measures will be, most probably, defeated. This is, perhaps, of all others the case in which extirpation of the part might be expected to be successful. But he does not recommend it, because the operation is a very formidable one, and, he says, this disease can be cured without it. Another reason why he rejects it is, we have no means of accurately determining whether the taint is thus isolated, or whether other parts are not already contaminated, so that we run the chance of only obtaining that equivocal triumph in which an operation is blazoned forth as being crowned with brilliant success, while the patient dies of the disease for which it was performed.

We will next give the theory and practice of Dr. W. H. WALSHE.

Dr. WALSHE considers cancer no longer as an external surgical malady, but as a true constitutional disease, of which the local manifestation is a small and often an unimportant feature. Dr. W. puts the diagnosis of this disease, when it affects internal organs, in a much clearer light than it has been presented heretofore. The writer observes, that the strongest proof of the constitutional nature of cancer is the fact established by Dr. WALSHE, that the knife should only be had recourse to in the extirpation of the local affection in a very small proportion of cases. Dr. W. commences his views on the nature and treatment of cancer by saying: I had scarcely commenced the study of adventitious products before I became convinced that much of the obscurity pervading the subject arose less from its nature than from the erroneous manner in which its investigation had been conducted. I found observers had overlooked the fact that the higher orders of these products were real existences developed within existences, and possessed of two distinct modes of life—a life subject to its variations of health and disease, irrespective of the organism in which they had taken birth, and a life influenced by the various structures and functions of that organism.

He saw that phenomena, accessory and contingent, were confounded with others necessary and essential, and that, as a natural consequence, misapprehension of many pathological relations had followed. Desirous of removing this source of unsound doctrine as it affects the most important of adventitious growths, cancer, he separately considered the healthy and diseased conditions of vitality of the product. The following is his view of the nature and sequence of this disease: "A certain constitutional state exists, and may continue to exist for a variable period, without giving functional evidence of its presence, although the blood and solids of the body are specially modified. In consequence of local injury, or otherwise, exudation takes place. Upon that exudation the constitutional states have impressed special attributes and tendencies: among these attributes rank an intrinsic power of vegetation. This vegetating faculty of the exudation reacts on the system by constantly draining it of a portion of its nutrient materials. The progeny feeds upon the parent organism, and the first phases of evolution are accomplished, but the natural tissues have been so modified in properties by the constitutional state, that they are incapable of resisting the encroachments of the vegetating exudation, and hence become the seat of atrophous, ulcerative and other modes of destruction. Discharges of various kinds now drain the system of its fluids, and impair its vital energies; and the second phase is established. Meanwhile, secondary alteration of the blood is effected. This fluid becomes the vehicle for the circulation through the system of elements possessed of a germinating force. These stagnate, are deposited, and new local vegetations spring into life and activity. The same series of phenomena are again and again gone through, until the system, drained of its reparative fluids in feeding exudations and supplying discharges, is exhausted of almost every drop of pure blood through the influence of secondary cancerous impregnation. Paralyzed in its nervous energies by physical anguish and deficiency of pabulum, it sinks in the struggle against the superior powers of the new existence it has created, and in death has closed the third phase of this disease."

Here we find Dr. WALSHE to stop short after giving this lucid and learned view of the pathology and termination of the disease, without giving us any remedy for its arrest or cure. But he proceeds to give us the treatment of several other physicians, and first, of Dr. ARNOT's method of treatment by compression. In the year 1809, Dr. SAMUEL YOUNG conceived and acted upon the idea that the continued nutrition of scirrhous tumors might be completely prevented, and the absorption of existing substance insured, by submitting them to methodic compression. The results of the practice, as made public by himself, are as follows:

The number of cases is nineteen; of these, seventeen relate to cancer of the breast, two to ulcers of the cheek and upper lip. Twelve cases terminated by cure; five were considerably benefitted. The two cutaneous ulcers improved somewhat. The majority of the tumors were hard, irregularly tuberculated, and the seat of lancinating pain. Six of them were ulcerated, and discharged ichorous pus; even in the worst cases the tumor diminished in size; but the patients fell victims to the diathesis. In consequence of Mr. YOUNG's announcement, the plan was tried at the Middlesex Hospital, and a committee appointed to report upon the results.

The conclusion of the report, drawn up by Sir CHARLES BELL, was that compression could not be regarded as a specific cure, and had no claims to notice, except for its power of alleviating pain. But as was justly rejoined by Mr. YOUNG, much of the want of success described may have arisen from defective management of the plan. No details of the cases are given. This was, to say the least of it, not fair. The spirit in which Sir CHARLES BELL judged, may be inferred from the allusion to the mode of treatment as a specific; a character, with which Mr. YOUNG never sought, in the remotest degree, to invest it. It would in truth, be just as wise, observes the latter, to speak of the pad of a hernial truss as a specific against strangulation, as to assign this character to compression in cancer. The testimony of Mr. TRAVERS is favorable to the practice. He has known tumors such as those already described, gradually reduced, and at length absorbed by equal and persevering compression, as by strips of soap and adhesive plaster, or by an elastic roller passed many times around the chest, with layers of amadou interspersed between the folds of the roller.

M. RECAMIER has employed compression upon a very large scale, and the more important parts of his results are as follows:—of one hundred cancerous patients, sixteen appeared to be incurable, and only underwent a paliative treatment; thirty were completely cured by compression alone, and twenty-one derived considerable benefit from it; fifteen were radically cured by extirpation alone, or chiefly by extirpation and compression combined, and six by compression and cauterization. In the twelve remaining cases, the disease resisted all the means employed.

MM. BLIZARD and MUSSAN, have published three cases, and M. CARRAN DU VILLARDS three others; in all of which, irregular nodular scirrhi, the seat of lancinating pain, etc., were removed by compression. Dr. A. L. I. BAYLES gives as the general results in 127 recorded cases, 71 cured; 26 instances of improvement; 30 total failures. These results, the most favorable on the whole that can be adduced in favor of any mode of treatment (then known), bear scrutiny of the severest kind. It is no doubt true, that in some of the cases alleged to be cancerous, neither of the anatomical species of cancer existed; but it is on the other hand perfectly unquestionable that many of the absorbed growths were not only actually scirrhous, but had already become the seat of ulceration when submitted to compression. Difference of opinion has existed as to the best mode of applying compression. M. RECAMIER employs perfectly smooth disks of agaric, laid over each other and retained in situ by a roller as the compressing materials.

M. BEGIN sometimes substitutes a laminated piece of lead, modeled to the tumor, and surmounted with a pyramid of graduated compresses. This application (which is far from a novel one) frequently becomes painful, and cannot be endured. M. BEGIN recommends a renewal of the apparatus every day, or every other day; but he thinks it better to change only when the bandages become loose; and perhaps in consequence of this view, an elastic corset capable of accommodating itself to the size of the part, as the compressing agent, wherever circumstances admit of its use, would be best. But all contrivances of this kind are of difficult application for various reasons. In the first place they exercise uneven and irregular pressure on the tumor; in the second place they confine the movements of the chest, to a degree varying with the amount of contraction; in the third place, the force employed is not directed against the diseased mass alone, but wasted in a great measure upon the healthy parts; and in the fourth place, while the difficulty of applying the apparatus effectually is extreme, it invariably becomes loosened and more or less disarranged within a short period after its application. Besides all this, the least unevenness in the material lying next the diseased structure, renders the compression unbearable from the pain it produces. These are the chief reasons, doubtless, which have hitherto prevented compression from taking its place as a general system of treatment of various external cancers. In order that we may have all the light on this subject that we can command, we will give Dr. ARNOT's plan of using mechanical means in the treatment of cancer.

He has invented a method of applying compression, which, while it is free from all objections mentioned, is philosophical in principle, and possessed of peculiar practical excellence. His apparatus consists of a spring, an air cushion, supported by a flat resting frame or shield, a pad, and two bolts. The spring, which is steel, is the compressing agent, its strength being varied with the amount of pressure it may be desired to sustain. The shield varying in shape somewhat with the circumstances of particular cases, is generally slightly convex on the internal surface, of circular or oval outline, and formed of a ring of strong wire, connected at two opposite points by a flat piece of iron, which serves for the support of the springs, screws, etc., the whole being covered with jean. To the rim of this is sewn a conical cap of soft linen, designed to receive the air cushion, to keep it constantly slack, and prevent it from slipping about when applied. The air cushion thus kept slack, fashioned into a sort of double night cap, lying in a position within the inner surface of the shield, and sufficiently filled with air to prevent the latter from pressing directly on the part, receives within it the tumor to be compressed. One end of the spring is attached by screws to the external surface of the frame, and the other to the solid but soft pad placed wherever the counter pressure is to be made.

The straps are used to keep the apparatus steadily fixed. Let us suppose the breast is the region to which the apparatus is to be applied. The position of its various parts will appear manifest. The spring may be either passed over the shoulder or around the waist. The latter mode of application suits best when the tumor is seated toward the external border of the breast and inclined to slip toward the axilla. The mechanical advantages of this mode of compression are, that the movements of the thorax are not interfered with; that the amount of pressure may be regulated to a nicety; that the whole morbid mass undergoes constant equitable and uniform pressure; that the part is protected from external injury, a matter of great importance; and that except in a very few cases the apparatus may be very easily arranged. It is necessary that the amount of pressure should be low at first, not over two pounds, especially in the case of nervous irritable people: in fact, the instrument should rather supply a support for, than exercise pressure on the part; that the entire morbid structure should be enclosed within the cushion, and in all cases there be a distinct thickness of air cushion between the shield and the skin. The effects produced by pressure are removal of adhesions, cessation of pain, disappearance of swelling in the communicating lymphatic glands, gradual reduction of bulky masses to small hard, flat patches, or rounded nodules, which appear to be locally, and generally perfectly innoxious; and in most favorable cases totally to remove the morbid production.

The relief of the pain by this instrument is without exaggeration almost marvellous. This effect is insured by the peculiar softness and adaptation of the air cushion: the medium through which the pressure of the skin is transmitted to the surface.

Females unable to obtain sleep even from enormous doses of laudanum, cease to suffer on its application; and sleep, therefore, as if they were perfectly free from the disease. This remedy, however, is not adapted to every case. There are certain conditions which either interfere altogether with the use of this instrument, or reduce it to a mere palliative agent. These conditions are more particularly excessive bulk of the new growth, and such localization of this structure as to place any portion of it beyond the reach of pressure. Persons of irritable skin and temperament, and prone to become edematous or anasarcous are managed with some difficulty. Less is to be expected in cases of encephaloid cancer than in those of other species; and in cases of infiltrated than in tuberculous accumulations. If the morbid mass be extensively softened, ulcerated, or in a state of fungus vegetation, palliation is all that can be fairly hoped for. Adhesion to the skin is of untoward influence. As a means of controlling and averting hemorrhage, the slack air cushion is of high utility. Dr. Arnot says his air cushion is applicable to cancerous tumors in every situation, where a bony or other solid support exists behind the growth, and where a point of counter pressure can be had. The mamma, the limbs, the surface of the thorax, or cranium, are the seats in which this mode of treatment is most readily applicable. He says: " I see no reason why cancer of the testicle might not be treated thus; and gentle pressure on this plan in certain cases of cancer of internal parts, might relieve pain, provided the general functional relations of the parts do not interfere, (and they will not often do so) with the adoption of such pressure.

The system of pressure now described, useful as it is, independent of all other treatment, may be rendered more efficacious by the association of other external means, and internal remedies. The following case exemplifies the power of such combinations. I herewith give a case treated by pressure, by Dr. WALSHE. The reader can judge whether it was truly carcinomatous or not. He was requested by Mr. LANGSLEY, March 3, 1843, to see a lady affected with scirrhus of the breast, which it was proposed to remove by the knife. Exactly five months and a half ago, she was attacked by slight pain in the right breast, and found there a lump about the size of a small hazelnut, not tender to the touch, unattended with soreness or discoloration of the skin: it increased but little in size till the last six weeks, within which time it has enlarged to its present bulk. She suffered scarcely any pain in the tumor itself; but had lancinating pains above the nipple in the indurated gland, to be described presently. In its general outline the right breast is double the size of the left. It has always been somewhat the fuller of the two, the subcutaneous veins more visible than on the left side. At the axilla border of the gland, is an excessively hard, solid, defined, rather movable tumor; the finger may almost be slipped behind this; but at its inner edge it is continuous with another indurated mass. Obviously a portion of the mammary gland itself is in a state of infiltration. The tumor is finely knotted on the surface, and the infiltrated part somewhat enlarged. Beside this, the substance of the gland is indurated and knotty, especially about the nipple. The nipple is less prominent than on the healthy side; but is not actually drawn in. The areola is unaffected. There is no adhesion of the skin, or alteration of its texture; no enlarged glands in the axillæ; but a slight thickening and hardening of some of the absorbents leading thereto; no discharge of blood even from the nipple. When the tumor was examined, it was not painful, but became so a short time afterward.

Measurements.—Whole breast four and three-quarter inches broad, three and a quarter inches vertically; tumor one and three-quarter inches broad, and one and a half inches vertically. The soft parts about are flaccid and yielding. He gave the following pill:

R.—Arsenici ioduret.,   - - - gr. j.
Extracti conii,   - - - - ʒij.   M.
Et in pil. no. xvj div. Signa.—One to be taken twice a day. Diet, light and nutritious; moderate walking exercise.

March 8th. Applied the slack air cushion. Diameter of the bag six and a quarter inches; pressure of spring, three and a quarter pounds. The whole mass of the breast was included, except about half an inch at the left superior angle, where merely cellular tissues remained uncovered. No annoyance of any kind was experienced by the patient, except slight impediment in respiration, which ceased in a few minutes.

March 10th. Pain totally removed, size of general mass of breast somewhat diminished, but the tumor is only rendered more prominent and apparent by this. No inconvenience is experienced from the instrument except in the back; the patient being thin, the pad presses uncomfortably on the spine.

March 12th. Tumor appears very considerably reduced in bulk, more elastic, less stony in feeling, less pointedly knotty on surface, less sharply defined; all these changes are to a very small amount, but they are nevertheless positive. Catamenia for the last two days.

March 30th. The tumor which has been gradually decreasing is now scarcely more than the original size. Some slight indications of absorbents with tenderness of the skin and slight redness: nothing in axillæ. From this period until the middle of August, the progress of things was slow and interrupted. Twice at the menstrual period, the tumor enlarged slightly, without, however, becoming painful. He gradually increased the force of the spring (which had always been carried over the shoulder) to six pounds, and diminished the diameter of the bag to three and a half inches, by which a great increase of pressure was obtained; and at the date named, the tumor was about one-third only of its original size, and had become freely movable under the skin, and the general knottiness and hardness of the gland had almost disappeared. Patient had lost altogether her headaches which used to torment her. Here he lost sight of his patient, partly by his, and partly by her absence from the city till the middle of November. During all this time the instrument had been more or less neglected; and not applied at all for the last month. The use of the pills had also been interrupted.

November 3d, 1843. The tumor is now at least half as large again as when he first saw it. It is more painful than ever and had reacquired all its original characters: it is, however, still adherent. He reapplied the cushion of the diameter first used, and within a week, a favorable change had taken place; the tumor continued thenceforth to diminish in size, until it was reduced to the size of a hazelnut, this little nodule appearing immovable. He at the close of January, 1844, directed the iodide of lead ointment to be applied on the part daily, the pressure at the same time to be continued. The effect was almost immediate, so much so, that the patient after the lapse of a fortnight, requested to use the inunction without the pressure. The impulse to absorption had been given. The tumor gradually decreased in size, and had totally disappeared at the close of April, 1844. He saw his patient August, 1845; examined the breast, and found it in every respect like its fellow, without a vestige of tumor or induration of any kind. Dr. WALSHE remarks, here then was a tumor, which, though it had not given rise to any of the more terrible evils appertaining to cancer; yet possessed the elements and best defined manner, and the sum of characters assigned by universal experience, to growths pursuing the common course of that disease. This tumor disappeared completely under the persevering use of the means described. Had the growth been removed with the knife, the chances are extremely strong (from analogy) that within the present period, the disease would have reappeared, and perhaps would have destroyed its victim.

But from the very perfection of success in this and similar cases, an objection may arise in some minds. It may be urged, that as in such cases a mass composed of indefinitely vegetating cells, is removed from its original site by absorption, the displaced cells may, in some new abode germinate and flourish. But the objection is a fallacious one. The absorption effected in such cases, must occupy physiological grounds, and be considered of the kind he has termed unproductive; and clinical experience, so far as it has yet gone, corroborates in the non-reproduction of tumors thus dispersed. (Doubtful.

Dr. J. H. BENNETT, Professor of Medicine in the University of Edinburgh, in the chapter on the rational treatment of cancroid growths, considers — First, the means of retardation and resolution; secondly, the means of extirpation; and thirdly, the means of prevention. The growth of cells in animals and vegetables is favored by an elevated temperature, a proper supply of moisture, room for expansion, and certain localities; and on the other hand, is retarded by excessive cold, dryness, want of room, and unfavorable positions.

Now, as cancerous and cancroid tumors chiefly or entirely advance by the development and growth of cells, by placing the affected part under the circumstances unfavorable to cell development and growth, their progress may be retarded or arrested. The direct application of cold, therefore, when circumstances will permit, may be employed with this view. The cutting off the usual supply of fluids for a time, by the ligature of the arteries leading to the morbid growth, has also, in some cases been attended with good results. The apparatus invented by Dr. NEAL ARNOTT, or that more recently invented by Dr. JAMES ARNOTT, of Brighton, with which external cold and dryness may be combined, seem the best fitted for carrying this mode of treatment into effect.

The means of extirpation are first, the excision of the part; second, chemical agents, which destroy texture. It is well known that many surgeons, discouraged by the frequency of the recurrence of cancerous diseases, feel indisposed to remove them by operation: the more especially as they believe that in many cases it hastens the fatal termination.

Other surgeons, again, take a more favorable view of operative interference: and believing that they have excised growths undoubtedly cancerous, which have never returned, and that in other cases, though ultimately unsuccessful, they have relieved the patient from much suffering, and have thereby prolonged life, advocate having recourse to excision. One of the arguments used by those averse to excision, in most cases, that it is barbarous to subject the patient to an operation, often very painful when we have little hope of its being useful, is now, in a great measure, removed by the late discovery of the anæsthetic effects of ether and chloroform, especially the latter. No doubt the imperfect methods hitherto almost generally practiced of distinguishing between cancerous and cancroid growths, accounts in a great measure for these discrepant conclusions of practical men.

Dr. BENNETT, after alluding to the successful results that have in some cases followed the excision of morbid growths, proved by the microscope to be cancerous, expresses himself in the following terms:—In like manner, by operating at an early period in all cases of suspected tumor, and keeping careful records, both of the minute structure of the parts removed, and of the ultimate results, much advantage would be gained to surgery. Lastly, by boldly excising all cancerous growths within his reach, when after careful investigation the surgeon has satisfied himself that no internal organ is effected, and repeating the operation so long as the return of it is merely local, he feels persuaded that not only in many cases would life be prolonged and much suffering saved, but some might be permanently cured. If this applies to cancerous, it does tenfold to cancroid growths, which every thing we know warrants us in asserting are much less fatal and malignant. In discussing the means of prevention, after pointing out how little confidence is to be placed in the internal remedies which have been recommended for the correction of that unknown predisposition in particular individuals to cancerous exudation, he throws out some general views for regulating constitutional treatment of cancer founded upon the difference between tubercular and cancerous deposits; which he however admits are purely hypothetical.

As VOGEL states, carcinomatous structures are distinguished from tubercular by their higher organization.

Dr. BENNETT argues that in proportion as the power of cell growth increases in cancerous growths, they abound more and more—in fact this excessive cell development must be materially modified by determining the amount of fatty elements which originally furnish elemental granules and nuclei; and that a tendency to the formation of fat would seem *a priori* to be opposed to the cancerous tendency. If a tendency to fat be an antidote to tubercle, as he believes it is, spareness may possibly be considered opposed to cancer. In the one case we should do all we can to bring the nutrition up to, and above the average, in the other down to, and below it. This plan of treatment does not preclude the endeavor to invigorate the general health by exercise and attention to the secretions and excretions. We might not have noticed such prophylactic suggestions from their being so extremely hypothetical, were it not that our deplorable ignorance regarding the constitutional treatment, both empirical and rational, of cancerous affections, induces us to listen patiently to speculations regarding it, which in other circumstances we would not be inclined to tolerate. The profession is much indebted to Dr. Bennett for his valuable contribution to our knowledge of the microscopical structure of cancerous and cancroid growths. It bears ample evidence of his great zeal, industry and success in pathological researches. (We shall have further use for his remarks on tubercles and cells.)

We extract the following analogies of some points in the publications of Drs. Walshe and Bennett made in 1849, July—on cancer. The following is the description by Dr. Bennett: "The microscopic appearances of a specimen of true cancer of the mamma. A thin section was removed from the centre of the tumor, and examined by a power of 250 diameters linear. It consisted of a meshwork of fibrous tissue forming waved bands arranged here and there in circles varying in size; some of these were very large,—one-fifth of a millimeter in diameter, inclosing other circles, each of which was surrounded by several filaments of fibrous tissue. Some of these circles contained numerous nucleated corpuscles crowded together, mixed with granules; others contained only a few or nothing but granules. Here and there were to be seen several compound granular corpuscles. On adding acetic acid to the whole structure, it became more transparent. Many of the fibrous filaments became invisible, and such as remained were studded here and there with elongated nuclei. The walls of the corpuscles were partially dissolved and rendered very transparent while their nuclei were unaffected. The cream-like fluid which could be squeezed from the surface of the tumor, contained faint nucleated cells; second—compound granular corpuscles; third—numerous granules. The nucleated cells were of a round or oval form, varying in the largest diameter from the 1-100th to the 1-50th or even the 1-40th of a millimeter in diameter. Some contained one nucleus, others two of an oval form, varying in the longest diameter from 1-100th to 1-75th of a millimeter. Some of these nuclei contained one nucleolus and others two. On the addition of acetic acid, the cell walls were rendered more transparent: the nuclei were unaffected, and appeared in consequence more distinct." This examination presents, according to Dr. BENNETT, a good example of the true cancerous structure. (I believe a better never will be found.) He selects this case as the standard of comparison, and the cells he describes, he designates as the true cancer cells.

He gives us another microscopic examination of another tumor of the breast. In this case the tumor was of great size, very hard and painful, the nipple unretracted, and the axillary gland unaffected. The physical appearance of the tumor was that of an amber-colored gelatinous mass, passing into an opaque, white, firmer, fibrous mass. The amber-colored gelatinous portion was composed principally of a structureless blastema, containing here and there fibres of extreme delicacy which with careful management of the light, might be observed to assume the form of corpuscles, in the midst of the jelly-like blastema. At different depths were a number of compound granular masses. Some of these when brought into focus presented a number of granules varying in size from the 1-500th to the 1-200th of a millimeter in diameter, which highly refracted light, and were aggregated together without being inclosed in a cell wall. When out of focus, these masses presented a dark-brownish shadow, where the amber-colored jelly was passing into the white substance. The fusiform cells became more numerous, and were mingled with a number of oval nucleated cells of great delicacy, varying in size and especially in length, some being caudate, others spindle-shaped. A thin section of the denser part of the white portion presented a fibrous structure, wholly composed of fusiform cells, which, on the addition of acetic acid became more transparent, whilst the nuclei were rendered very distinct. This tumor seemed to consist of a simple blastema in which fibrous tissue was forming. It contained no cancer cells, but apparently only the elements of exudation. Let us examine how far the observations of Dr. Bennett will enable us to determine more satisfactorily than heretofore.

First. How to recognize cancer before and after death.
Second. The physiological and pathological history of cancer.
Third. His ideas as to the mode of treatment.

The Doctor considers that the common subjective and objective symptoms, such as lancinating pains, unequal surface, hardness, elastic feel, ulceration, affection of surrounding parts, constitutional cachexia and return of the growth after excision, are symptoms which are not peculiar to cancer, being occasionally absent in tumors which are undoubtedly cancerous, and being present sometimes in epidermic fibrous growths of the most innocent nature. Accordingly he considers that an accurate microscopical examination, taken jointly with such evidence as may be afforded by the symptoms and progress of the case, is absolutely necessary before a correct diagnosis of the case can be given.

What then are the microscopic characters of cancerous as distinguished from cancroid growths? Dr. Bennett's opinion may be given in a few words: "If a tumor presents among other elements certain cells of definite aspect, and is at the same time situated in parts which render it possible that these cells can be epithelial, epidermic, or cartilage cells, that tumor must be cancerous." He gives the following account of their characters: "The cancer cells are of variable shape, round, oval, caudate, heart or spindle-shaped, etc. Of variable size, from the 1-10th to the 1-100th of a millimeter, nucleated, single, double, triple or more: but generally nucleated, colorless or with melanic deposit; increasing commonly on an endogenous plan. And thus after presenting the phenomena of parent cells, occasionally increasing by division of the nucleus, but never by division of the cell-wall or by splitting up into segments of the mother cell. The cell-walls are dissolved by acetic acid, the nuclei corrugated, and their margins thickened. Cancer cells never pass into fibres; they become caudate or throw out pointed prolongations, but the transformation proceeds no further. They are contained in great abundance in the white milky colored juice which can be squeezed from most cancerous tumors, and they are surrounded by a greater or less quantity of viscous fluid. Cancer cells cannot be distinguished from young plastic epithelial cells—ambrional and cartilage cells." In this description Dr. BENNETT agrees very closely with the account of the same structure given by LE BERT, WALSHE and HANOVER.

The chief differences are:—1. LE BERT does seem disposed to admit that cancer-cell can be confounded with any other species of cell; 2. WALSHE states that acetic acid does not dissolve the cancer cell wall, but merely renders it transparent; and that it can be restored by using ioduretted solution of iodide of potassium; 3. That Dr. BENNETT seems inclined to doubt the exogenous mode of formation of cancer cells, although the comparatively few cells of scirrhus are said by DR. WALSHE to increase mainly in this way. Disregarding, however, their differences, and assuming that the character of cancer cell given above are correctly detailed, can we by means of these microscopic characters distinguish cancerous from other tumors? As we have already mentioned, some normal cells, as of young epithelial and cartilage, are so like cancer that they cannot be distinguished from each other, and Dr. BENNETT says they cannot always be separated.

Thus in a case of cancer of the cheek, a milky fluid containing cancer cells was found in the flocculi formed by the hypertrophic curled elastic filaments of the dermis, and in the meshes of a fibrous matrix situated beneath the skin. In this case, Dr. BENNETT says, the epidermic scales afforded a striking resemblance to some forms of cancer cells. I shall hereafter point out how epidermic and epithelial cells, after being steeped in water or a thin corous fluid, may exactly resemble, in every essential particular, cancer cells.

Again, he says that an enchondromatous tumor of the ischium and pubis was considered by all who saw it to be cancerous. It had partially softened, the softened pulpy matter was contained in cells round, angular or caudate, with nuclei and nucleoli. If, however, a tumor is found at a distance from skin, mucous membrane or cartilage, and yet contain cancer-cells, then Dr. BENNETT seems to consider their presence as absolutely diagnostic. But admitting this, it seems unfortunately certain that a tumor may be truly cancerous—that is, may be connected with a peculiar constitutional taint, may be intractable and disposed to run through a certain series of changes—and yet no distinctive cancer cells can be detected in its structure. Without entering into an analysis of the numerous discussions on this subject, we will simply give Dr. WALSHE's statement, which is of very high authority. He says: We have known growths which have destroyed life, with the cachexia of cancerous disease, and which clearly exhibited the local progress and naked eye characteristics of encephaloid growths, which, nevertheless, were composed of non-nucleated cells, undistinguishable from those of common exudation matter. And in an after part of the same article the following series of propositions are given:—

Parent cells, containing within them sub-cells, having darker nuclei, and these in turn bright nuclei, are strongly characteristic of cancer. But such cells arise in, and may be altogether absent from, scirrhous encephaloid; in some phases of its growth may also be without them.

A tumor may present to the naked eye the characters of encephaloid, be the seat of intestinal hemorrhage, affecting the communicating lymphatic glands, run in all respects the course of cancer, and nevertheless contain no cells but such as are undistinguishable. In the present state of knowledge, from common exudation cells, the shapelessly caudate cell seems significant of cancer, but it may be absent from encephaloid, and is very rare in scirrhus or colloid.

Nay, more; while a primary malignant tumor contains these cells alone, the lymphatic glands secondarily affected may contain compound nucleated cells, spherically and shapelessly caudate. The true fusiform cell is an adventitious formation when it occurs in cancer, and has no special diagnostic signification, although we are of the opinion that there is no one microscopic character which will serve as an infallible rule or mark of cancer in all cases. We are disposed to agree with Dr. BENNETT in attributing the highest diagnostic value to cancer cells when they are present and free from ambiguity. They will be present in a vast majority of cancerous growths. Still we cannot admit that their absence negatives the idea of cancer. The microscope should be used in every instance where it is practicable in the diagnosis of cancer, especially where doubt exists as to its real nature.

M. KUSS, of Strasburg, has proposed to extract a portion of a doubtful tumor with an exploring needle, and to examine microscopically the portion thus obtained. In the case of open sores a microscopic investigation of the discharge will often clear up all doubts as to the cancerous nature of the disease, by detecting at once the characteristic cancer cells. Admitting that this microscopic examination of a tumor is of great value in the majority of cases, but may yet in some instances be insufficient to decide the question of its nature, are there any other circumstances which, apart from the more ordinary symptoms of cancer, may serve as diagnostic marks? Cancerous tumors almost always yield—when they are open—a milky fluid or albuminous juice on pressure. Dr. WALSHE thinks this is less likely to deceive us than any other, outside of the true cancer cell. Occasionally, however, no fluid can be expressed, as in the dry epithelioma of the nose or cheek.

One of Dr. BENNETT's cases of cancer in the mouth gave no fluid on pressure, although numerous true cancer cells could be obtained by scraping. And in some rare cases enchondroma gives a milky fluid by pressure. But there is another very important character belonging to cancer, and that is the property of infiltrating the tissues among which it lies. It extends its meshes in all directions, and in some cases, as in the liver, produces a transformation of the adjoining tissues into carcinomatous matter. Clinically, this character is of importance; though cancer is not always so infiltrating, but is often abruptly circumscribed, or even in some rare cases encysted. And even if infiltrated, this cannot always be recognized during life; but nosologically this is a characteristic of great importance. Dr. Bennett and Dr. Walshe both attribute great importance to this peculiarity.

Yet a doubt may be raised whether this infiltrating property is limited to cancerous growths; whether it may not belong to fibrous tumors, and after all the observations made by scientific men on this subject, may it not be, that exudation matter has been mistaken for fibrous infiltration, and the doctrine of infiltration be confined to true cancerous tumors.

It has been asserted also, that it is a peculiarity of cancerous tumors to return after extirpation, and there is little doubt that there is far greater liability to return after extirpation than in growths of a simple nature. There is an opinion, for some years gaining ground, that fibrous epithelial, and even fatty enchondromatous growths return after excision much more frequently than is generally believed. From all the evidence presented, it would appear that there is no positive and invariable evidence which in every case will fully satisfy us that the case is certainly cancerous; that the microscope does not always decide this vexed question, although the occurrence of diagnostic cell-forms is extremely frequent; and the milky juice is not always present; and infiltration, though generally, is not in every case present.

The physiological and pathological history of cancer, compiled from several eminent authors, is about as follows. We find many things therein with which we most heartily accord, and some things we shall dissent from when we come to give our individual opinion, whether they ultimately, in the progress of knowledge, prove to be right or wrong. In Dr. Walshe's great work on Cancer, he says, the first step in cancerous formation is an exudation of a peculiar blastema, which, in our present knowledge, can be distinguished from common nutritive blastema only by the difference in the resulting organized forms. Such a blastema, as a general rule, exudes completely from the vessels, seldom is arrested in the thickness of their walls, and very rarely, indeed, if ever, remains within the vessels. This cancerous blastema thus derived from the blood itself, is in an abnormal state when it can furnish so unhealthy a substitute for the ordinary secretive and nutritive blastema. And this inference is corroborated by almost all the facts we can bring to bear upon this point, and which all tend to the conclusion that a cancerous tumor, under all circumstances, is but a local evidence of general vitiation of the system. Subsequently the cancerous deposit may remain quiescent, the cancerous cachexia which produced it having passed away, or may advance with greater or less rapidity, according to the tendency of this cancerous diathesis. If it proceed, it is destructive, or malignant: partly in proportion to the degree of this preceding constitutional disease, and partly in proportion to the infiltrating power possessed by the cancer tumor in the particular individual. Finally, the local disease itself reacts on the constitution, and impresses a deeper and more advanced influence on the general system.

These views are identical with those held by Dr. BENNETT. He says, the great majority of facts with which we are acquainted, lead to the conclusion that the filaments, cells, and fluids, which together make up the tissue I have called cancerous, originate in coagulated exudation. This is poured out exactly in the same manner as all other exudations, viz., by enlargement of the capillaries, their repletion with blood, and the transudation through their coats of the transparent liquor sanguinis, which, coagulating outside the vessel, constitutes an exudation; which must differ from the exuded matter in what is called inflammation or tubercle, but in what that difference consists he does not say. The characters given to the blood by ROKITANSKY and ENGAL to distinguish what they have called the cancerous dyscrasia, are so vague and uncertain as to have no real value; neither is there any proof that such dyscrasia consists in an excess of albumen or fibrin in the blood. The latter takes place after all exudations. But although the present state of pathology does not warrant our stating positively wherein exudations differ, we cannot avoid seeing that their occurrence in different individuals produces very different results. Thus, if an exudation be poured out in a healthy individual, it produces a series of phenomena, which we call inflammatory; if in a scrofulous individual, another series of changes, called scrofulous, or tubercular; whilst in a third person, the result may be cancerous growths. Every kind of reasoning must lead us to the conclusion that these different changes and effects depend not upon the vascular system, which is the mere apparatus for the production of exudation; not upon the nervous system, which leads to the necessary derangement of that apparatus; and not on the texture which is the seat of the exudation, as that varies, whilst the cancerous formation is the same, but on the inherent composition or constitution of the exudation itself. Here pathologists pause, and tracing back to the blood, they are content. They have not sufficiently taken into consideration that blood itself is dependent for its constitution on the results of the primary digestion, in the alimentary canal, on the one hand, and the secondary digestion in the tissues, on the other. Pursuing these views still further, a comparison between the characteristics of the cancerous, and what may be termed the simple, or inflammatory, and the tubercular exudations, may lead us to some useful hints respecting the signification of each. Let us suppose that an individual in perfect health has from any cause an injury or irritation leading to inflammation, etc., an exudation poured out into any part, which is plus the normal exudation required for the repair of the tissues. The further changes which occur in this vary infinitely according to the part, the degree of irritation, the rapidity or slowness with which the blastema is poured forth, etc.; but one or other of what may be termed the healthy or normal changes occur. The exudation becomes the medium of adhesions, or is organized into false membrane, or accumulates in the interstices of organs, assuming a greater or less resemblance to them, and impeding more or less their functions; or it breaks down more or less rapidly into evanescent cell-forms which are discharged at once from the free surfaces, or are eliminated from the body through the medium of further changes, and different emunctories. The composition of this blastema is no doubt within the healthy limits; and these variations are evidently not dependent, for the most part, on any local conditions of the part in which the blastema is exuded, but are attributable to the general condition of the organism; to the purity and excellency of the blood, which is the indication of the true performance of all the varied metamorphoses of the frame. But instead of this healthy constitution, let the individual be tuberculous, or cancerous; then in certain organs, the exudation emitted from any cause, does not pass through this series of normal changes, but assumes one or other of the forms peculiar to the diathesis. Then in young subjects, and in persons of peculiar and defective assimilating powers, the exudation is below the healthy standard; it attains imperfectly any cell-forms, and is liable to rapid disintegration, and complete loss of vitality.

This is the tubercular exudation, and is poured out in its characteristic form, only on certain organs and positions. We see enough to convince ourselves, however, that a blastema poured out in other situations, as in accidental breaches of continuity, although it may not possess the tubercular character, is yet often evidently different in some way from the reparative blastema, which would be effused in a perfectly healthy condition of the system. In an older person, in whom the assimilating functions are also deranged, although in a different manner, the blastema poured out may be entirely dissimilar to the tuberculous blastema, and may have the tendency to the production of very highly-developed and reproductive forms, which grow with great energy and rapidity, and retrograde and disintegrate slowly and imperfectly. Between this diathesis and the former there is, therefore, almost an antagonism. They can hardly exist in the same person, and the presence of the one may be said almost necessarily to exclude the other. Then the hypothesis we have given above,—for it is not much more,—agrees with the facts of the case; for it is well known that tubercle and cancer attack at different periods of life, and affect different organs; while, when recent, they are rarely found associated. The word destructive, if applied to these tumors, would probably explain their nature and tendencies better than the one usually employed. They are destructive not only by destroying the tissues in their immediate neighborhood, but also by the new tissues deposited being peculiarly liable to decay, causing breaches or chasms, which may increase to a very great extent.

These, then, are the leading features of these growths: first, degeneration of the natural tissues around them; and, secondly, the tendency to decay of the tumors themselves. There is no doubt in the investigation of their growth. The microscope is a very valuable assistant to us here. Nevertheless, the many difficulties and sources of error necessarily intertwined with the use of this instrument, should warn us against the too implicit reliance which has hitherto been placed upon it in morbid anatomy. It cannot be relied upon, unless the observer has bestowed much attention and time to its use. Time may be regarded as a sure test; but at present we should not rely upon it to the exclusion of all other sources of knowledge. It is now commonly argued that the greater number of cancerous growths consist of a fibrous nature, disposed in meshes, which contain nucleated cells, nuclei, and granules, together with a thin serous or gelatinous fluid. The proportions in which these three elements are combined vary greatly in the different species of cancer. In the scirrhous, or hard cancer, (fibroid,) the fibroid structure predominates; in the encephaloid, or soft cancer, (fungoid,) the cells are more abundant; and the colloid, (hematoid,) or gelatiniform cancer is in great measure a jelly-like fluid. Neither of these elements, examined separately, are found to exhibit any decidedly distinctive characters; the peculiarity of the cancerous growth consisting not so much in the materials of which it is composed as in the mode of their arrangement, their location, and development. The fibroid basis of cancer seems to resemble clearly the common areolar or fibrous tissue; but this fibrous basis does not possess so distinct a fibrous structure. The cell element of cancer has naturally proved an object of much interest, inasmuch as the essence of the disease would seem to be almost concentrated in it. The rapid growth, and self-destructive or decaying tendencies of cancer, being to a very great extent proportionate to, if not dependent on, the quantity and location of cell development, viz., the meshes of the fibrous stroma, pathologists hoped, therefore, to find some distinctive structural peculiarities attached to these cells, whereby the nature of this disease might be recognized at once, and with certainty.

There is no writer, as yet, who has confirmed this by observation. But it is the opinion of those best qualified to judge, that cancer cells have some peculiarities which distinguish them from common pus cells and from tubercles. In their primary forms they vary in appearance a good deal, and in each of these forms they resemble some of the primary cells of other tissues. And if the form which is of frequent occurrence in specimens of well-marked malignant disease may be taken as the type, they are not to be distinguished from the cells of cartilage, or the cells composing the deeper layers of epithelium. Indeed, some of the structures which present, in a well-marked manner, the practical tokens of malignancy, do not appear to contain the cancer cell as an ordinary constituent. Such are some of the varieties of cancer of the skin, and of the muscular coats of the intestinal canal.

The cells commonly found in cancers, which may be seen in the fluid scraped from a section of this morbid mass, and to which the name of cancer cell has been applied, are larger than blood globules, of about the same size, or rather larger than pus globules, composed of a tolerably well-defined cell wall, with one, two, or more nuclei, in which a nucleolus may be often seen. The cell wall is rendered transparent, or dissolved by acetic acid, the nuclei being unaffected. The cells vary a good deal in shape, being round or oval, or elongated at one or both ends into a caudal or spindle form; or they may send out processes from various points of their circumference. It is the general opinion that they do not, at least ordinarily, advance beyond the cell stage of organization; that they are not transitional to any other form of structure, but that, having attained to a certain size, they are disintegrated, and their constituent atoms are returned to the fluid form from which they were derived. By a process of endogenous growth, they may become the parents of other secondary cells, found in their interior; that is to say, their nuclei and nucleoli enlarging, may be converted into cells, and fill up the interior of the parent cell, and at the period of its dissolution may be set free, and undergo like changes in their turn. The materials for the growth of the nuclei being furnished by imbibition through their walls, some authors have said their cells increase by fissiparous germination; and Dr. BENNETT speaks much of their occasionally undergoing a retrograde process: the nuclei disappearing instead of enlarging, the cell walls thickening, collapsing, and becoming infiltrated, or intermixed with oil, and ultimately dissolved into fragments or granules. He finds the yellowish masses, often found in cancers, to consist in a great measure of cells thus formed.

With regard to the nature of these cancer cells, and their physiological relations to the fibrous element and the natural tissues of the part affected, it is rather difficult to form an opinion. By some authors they have been looked upon as an independent existence, like entozoa or fungi, growing in the body, and propagating themselves in distant parts through the medium of the blood, by sporules, which find their way into the circulating current, either through the delicate walls of the capilaries, or through openings in the small vessels. It seems to me most probable that they are the result of some peculiar influence exerted upon the nutritive elements of the blood, effused among the tissues by an altered or morbid state of the part, being in this respect, to a certain extent, analogous to pus cells; which they also resemble in their capacity to undergo further organization, and their being interminal, not transitional, stages of development. The two morbid products differ, however, in this important particular: the cancer cells possess the property of self-multiplication, absorbing and assimilating the elements of the blood, whereas the pus cells are dependent for their increase on the continuance of the inflammatory process, causing fresh exudations; indeed, with similar tendencies to those from which they sprang themselves.

There are some circumstances, such as the accumulation of pus in certain cavities, or abscesses, which render it probable that even the pus cells exert a certain assimilative influence upon the fluids, or even the solid lymph, effused in their immediate proximity; but they do this in only a slight degree, and with a force scarcely comparable to that of the cancer cells. This point of difference between the products of inflammation and the elements of cancer, is of the greatest importance, and constitutes one of the marked features of their distinction. Let this not be forgotten.

Besides the cells just mentioned, there are commonly found floating in the same fluid nuclei and granules, with perhaps other cells, such as those which are in process of development in the tissues, pus cells, etc. Oleaginous matter in the form of minute globules dispersed through the mass, is a very usual constituent; and there may be found many elements, not belonging to the cancer itself but to the structures in which it is produced, such as striated muscular fibre, fat, granular tissue, etc. The blood-vessels are often of large size, and very numerous. They are probably derived from an increased growth of the vessels naturally supplying the affected part. It should be observed here of the two chief elements of cancer, that they are not only intermixed in very different proportions, but they do not invariably co-exist. In some few instances, the cells are present, being infiltrated among the tissues of the affected organs, or they may be intermixed with lymph, effused in consequence of inflammation taking place in the person who is the subject of malignant disease, or in whom the cancerous diathesis is strong. So, on the other hand, cases now and then occur in which a tumor, consisting only of fibrous stroma, without the intermixture of cancer cells, and resembling, therefore, a fibrous tumor, may exhibit the destructive and inveterate qualities of malignant disease. As a general rule, the rapidity of growth, and the tendency to enlarge in a tumor, are proportionate to the quantity of cells developed in it. They are greater in hematoid, less marked in fibroid, more abundant in fungoid, and least of all in epithelial. In the fibroid they constitute almost the entire mass

The relation which the cancerous elements bear to the natural tissues, and their effects upon them, are points of so much interest, that we again revert to them for a moment. It is evident, that at first the natural and morbid structures are, to say the least, very closely connected together. Most pathologists agree, that the cancerous matter is infiltrated amongst the original elementary parts of the parent tissue, and occupies nearly all the interstices, and that in process of time the elements of the tissues become compressed, appear to be blended with the deposit in a homogeneous mass, and gradually become atrophied and disappear, but this is very rarely the case. Indeed the connection between the natural and morbid structures is in some cases so intimate, the one being scarcely distinguishable from the other, as necessarily to suggest the idea that there may have been, not merely a blending, but an actual transformation of the healthy, into the cancerous tissues. This is particularly true in some of the glandular organs; in the liver, for instance, resulting from absorption and deposit, or from a primary formation. The first indication of the disease is commonly afforded by a mottled appearance resulting from a slight discoloration of some of the lobules; they retain their proper shape and size, and their distinctness from the neighboring lobules, but they are of white color, and the natural components of the lobules are replaced by, if not transformed into, the elements of cancer. As the disease increases, adjacent lobules are affected, their structure is more completely altered, their individuality lost, and they are fused into one mass. When once the cancerous change has commenced, it is almost sure to spread. The assimilative energies of the morbid products are so intense, that the neighboring tissues yield under its superior force, their own native powers, and are, as it were, prostrated, and soon disappear in the same manner as the part first affected.

Before being thus completely destroyed, they are found in many instances to undergo certain alterations, such as atrophy, or fatty degeneration, which indicate that their nutrition is impaired, and that they are ready to fall an easy prey to the destructive influence which is encroaching upon them. Whether these alterations are the entire results of that influence, operating upon them, though at a distance, or whether they are, in part also, the result of some inherent deficiency, which is the common cause, both of the cancer and the atrophy, I will not pretend to say, but am inclined to think the latter is most likely the cause. At any rate it is not uncommon to find bones affected with cancer in some parts, and atrophied and greasy in others. The mammary gland, which is the seat of scirrhus, in some of its lobules, is often shrunken or loaded with fat in others. Sometimes you see a small scirrhous lump under the nipple embedded in a mass of fat, which occupies that seat, and retains the shape of the mammary gland. In like manner, when a nucleus is affected with cancer, the fibres contiguous to the morbid product, are often observed to be in a state of more or less advanced degeneration; their transverse markings are indistinct, or invisible; their nuclei imperfect, and they are pervaded with oil. Further, although the general wasting of the body, so constantly accompanying the progress of the cancer may be occasioned by pain, discharges, and various accidental causes, the fat both under the skin and about the internal organs, having a remarkably deep yellow color, is attributable to some peculiar condition of the nutritive functions, dependent on, or associated with, cancer. It seems to be the general rule, that the higher or more organic tissues, the striped muscular tissues, and the glandular, are commonly affected with atrophy and degeneration, in connection with cancer, and that the more simple and less organized tissues, the unstratified muscular, the cellular, and fibrous, being less amenable to the destructive influence of cancer, are often hypertrophied, or thickened by its first impression. We frequently find the two effects simultaneous in the same organ: the glandular element wasting, becoming fatty, or replaced by cancer cells, while the fibrous element is increased to many times its natural thickness. We should not forget that the latter also does, after a time, yield to the destructive influences of the disease, being impregnated with cancer cells, or infected with the general tendency to decay and ulceration. When you cut open a cancerous tubercle you will often find that at or near the middle, it is softened or converted into a more or less diffluent pulp; and if you examine it more closely, it will seem to have lost all regular structure, and to be composed of fragments, which appear to be disorganized remnants of cell and tissue. The softening depends probably upon a failure of the nutritive powers, taking place to such an extent that the component atoms are no longer in those structural relations into which they had been thrown, and in which they had been maintained by the vital and nutritive forces of the growth. It commences naturally at the part first formed, viz., the centre of the tubercle, and is similar to the softening that takes place occasionally in simple tumors, or more frequently scrofulous deposits and tubercles, and is similar also to the softening and disintegration of the natural structures often induced by inflammation. Though occurring in cancer, for the most part as a regular or natural process, it may be induced by any cause, that prematurely impairs the vital energies, such as an attack of inflammation, or a blow; strange and paradoxical as it seems, that these same products should possess such irresistible assimilative influence over the surrounding tissues, and should be so unable to maintain their own existence, so liable to decay and dissolution.

The softening may commence at one or more points, and sometimes leads to the formation of one or more cavities in various parts of the mass, containing a turbid fluid which often looks like pus, and may contain pus corpuscules. Indeed, the process frequently resembles and is associated with suppuration, the smaller cavities uniting to form a larger one, which approaches the surface like an abscess, bursts through the skin and forms an ulcer. When the cancer is situated near the skin the softening commences on the superficial side, and takes place earlier than when it is deep seated. This breaking up of the morbid mass is usually attended with some inflammation of the surrounding parts with effusion of lymph uniting them together, and rendering them a more easy prey to the ravages of the disease. At the same time the adjacent lymphatic glands are commonly observed to become inflamed and to participate in the malady, if they have not before done so. It is by no means uncommon for inflammation in the surrounding tissues to precede the softening of the cancer; indeed it is sometimes cotemporaneous with the earliest appearance of the disease, and attends it through all its progress so assiduously that some pathologists have regarded cancer to be only one of the multiform results of inflammation. The cavity or ulcer commencing in the manner first described, continues to increase by the progressive softening and disintegration of the neighboring cancerous tissues, so that particle after particle is separated and forms a component in the discharge, the disease still advancing in the circumference while the dissolution of the mass is thus going on at the centre. In this way huge cancers are sometimes formed and large parts of the body destroyed. Sometimes the work of demolition proceeds at a greater rate, mortification succeeds to ulceration, and considerable masses are detached. Now and then, in consequence perhaps of a slight inflammatory attack, the whole mass appears to slough away and leave a clean surface behind; but the hopes of a cure thus excited are almost always doomed to be disappointed by the reappearance of the disease in the sides of the chasm from which the slough has disappeared.

There are some points of resemblance between cancer and scrofula which it may be necessary to point out, inasmuch as some writers think they are only a variety of the same disease. Cancer is regarded by some physiologists as a variety of scrofula, others imagine that the tubercle is a variety of cancer. But we think there are certainly sufficient differences to induce us to retain in pathology the distinctions at present considered by our most eminent physiologists to exist between them. True, there are some points in which cancer and scrofula resemble each other, inasmuch as both consist in the infiltration of a new exudation substance into the interstices of the textures of a part, that infiltration being attended with a wasting or degeneration of the original tissues; both have a tendency to soften and to ulcerate, causing a loss of substance in the affected organ and leading to more or less extensive destruction of the body; both seem to be closely associated with, or dependent upon, a want of energy of the nutritive forces of the body generally, as well as the affected organ in particular, so that the existence of either disease at any one part leaves us to apprehend its appearance in another; both affect all the organs and tissues of the body almost indiscriminately, and presenting nearly a uniform appearance in whatever region it is found; and the one is often so much like the other, that it is scarcely possible with the naked eye to distinguish between them.

A cancer differs from a scrofulous mass in its greater inherent energy of growth; its more complete organization which is widened by the formation of the peculiar cancer cell; in the fibres and blood vessels in its substance; its more complete independence of inflammation; its power of growth, of assimilating the adjacent fluids and tissues, and in its more determined progress. The scrofulous matter is composed chiefly of granules, hardly ever attaining even to the cell stage of organization, and is dependent for its increase not on its own assimilative power, but on fresh exudation, which is commonly the result of an inflammatory condition. It often falls into a quiescent state, and is absorbed; or if it softens and ulcerates, the disease does not generally go on spreading in the circumference, and the ulcer for the most part heals. Such important, and practical, and pathological differences are undoubtedly quite sufficient to justify the separation of these two kinds of morbid products into distinct classes.

That tubercle should be ranked with the last of these two classes, and that it is the form which scrofula ordinarily assumes in the internal organs, I have already expressed my opinion. True it is the form of scrofula most nearly approaching to cancer in its comparative independence of inflammation, and its greater fatality. But the tuberculous mass has no higher organization than the scrofulous; it has no cells, fibres or blood-vessels, except those which chance to remain in it from the undestroyed tissues into which it is infiltrated, and these diminish in number, and disappear after a time, so that they cannot be regarded as proper elements of the tubercle, but merely accidental constituents.

Tubercles have no power of assimilation or of increase: they are fatal on account of their numbers, not like cancer, in consequence of the energy of their growth. Each tubercle having softened, bursts and discharges itself, has done its worst; it leaves no spreading remnant behind, and left to itself will contract and heal, provided it is not maintained and enlarged by other tubercles, forming in its walls and bursting into its interior. The non-existence of any cells, or other elements possessing a capability of reproduction contrasts an important point of distinction between tubercle and cancer, connected, as it is with the inability of the former to increase in size, and to exert a destructive influence upon the surrounding tissues. A cancer grows by virtue of its own inherent force; a tubercle by something extraneous depositing fresh material upon it. It may, I grant, increase in size a little, but no more probably than is accounted for by the softening and imbibition of fluids and other changes going on in it. Moreover tubercle very rarely exists with cancer in the same individual, though it is a frequent occurrence with scrofula. Indeed the ages and classes of persons most liable to scrofula and tubercle are found to be rarely afflicted with cancer. And yet many considerations remain of great importance in pathology for the complete elucidation of cancerous growths. Certain parts of the process can be explained rationally, although not yet actually demonstrated, as the formation of a cancerous structure exuded from the blood vessels. These conditions appear to be independent of but to occur somewhat in the manner of those of an inflammatory nature, and pass through their accustomed stages of growth in persons who are the subjects of the most extensive cancerous infiltrations deposits, and infiltrations from the same blood. In ordinary nutrition, development and secretion, particular tissues exercise an undoubted power of election or appropriation of the various materials circulating through them; in some cases changing the arrangement of the elementary particles, if they do not actually form certain substances by their own immediate action; and so may an organ which has either become the seat of a cancerous deposition, or is predisposed to it, in some way unknown to us, exert its peculiar elective force upon the same blood, and supply other parts for the material of healthy tissue.

The following appears among the desiderata, for the perfection of our knowledge of the anatomy and pathology of cancerous growths, viz., to ascertain whether fibres exist in all cancers, and are really essential for the completion of the structure, the attendant phenomena of which shall be indistinguishable from cancers in general; and whether the fibrous tissue of cancerous formation bears any constant proportion, in its amount, to the fibrous character of the organs in which these are developed, and is regulated by the law of analogous formations. These inquiries cannot be answered otherwise, than by numerous and accurate microscopical examinations of the structure of organs so affected; and especially of cancers, which occur in parts that contain no fibrous tissue whatever in their normal state, and in such as contain so small a portion of it, that an increase in its quantity may readily be detected. Of late years there has been so much said on tubercle, in relation to consumption, scrofula and cancer, that it may be profitable to review for a moment the opinions of some of our best and most scientific writers on this subject. The word tubercle has been applied to so many morbid changes which differ from each other in their clinical as well as in their histological characters, that from originally denoting a small rounded body, it has been made to include any process which results in the breaking down of living tissues and the formation of cavities.

When a word thus becomes altered in meaning; when some use it only in a restricted sense; and others, almost as synonymous with phthisis, it must follow that confusion arises in discussion; and thus while pathologists are describing the same morbid process, they will look upon it as tuberculous, or not, according to their preconceived notions on the subject. The term tubercle is beginning now to be restricted to the gray granulations, a definite body, formed of oval, or rounded nuclei, and generally surrounding a vessel or bronchus. In cases of tubercular meningitis, it may readily be studied in the pia mater; and it is also to be seen scattered through both lungs, or in other parts of the body. In other cases, the course of the disease is not so rapid. The granulations may be found to be aggregated together, and in the most advanced portions, to have become softened in the centre, and to have formed cavities; while in other parts of the lungs, gray tubercles may be seen scattered about, and in various stages of transformation. Accompanying this pathological state, there is a distinct clinical history, which much resembles typhoid fever, and looks like that disease, having great daily variations of temperature. To the yellow or cheesy masses so often found in the lungs, Dr. CLARK does not assign the name tubercle. Under the microscope they are seen to consist of round cells, with nuclei and granular contents, and as they extend, nutrition being cut off from the central parts, they soften and break down.

The origin of this cheesy matter may be a pneumonic product, which has not been absorbed or cleared up, or a syphilitic deposit, or as a result of embolism or concretions in the bronchial tubes. This appearance occurring to Dr. CLARK, although found accompanying the gray granulations, may be met with independent of them.

In another class of cases a more chronic change occurs, and instead of a pneumonic product, a fibroid material results. The pleura is found much thickened, and proceeding inward are numerous septa, which divide the lungs into segments, and this change takes place in the perivascular, or peribronchial tissues. It consists either in a prolification of the connective tissue-cells, or in an exudation of an amorphous material into the air vessels themselves. This process is one that may last for many years, and although cavities may form they extend but slowly, as there is such a dense wall of fibroid induration around. It generally affects one lung, and the left more frequently than the right, and is often accompanied by an amyloid state of the liver or kidney. It is also to be met with in rheumatic or syphilitic individuals, or drunkards. Clinically, it is remarkable for its long duration, as the disease may last for many years. The absence of night sweats or any rise in temperature, may be accompanied by hemoptysis, diarrhœa and loss of flesh. Physical signs of cavities, or consolidation of lung tissue, are to be met with, although they may occur without any gray granulations being present. Yet this is not always the case, and if this fibroid induration surround tubercles, it may be looked upon as due to a slow process of irritation; which, beginning in congestion, ends in prolification of connective-tissue nuclei. While in other cases, where the congestion is more intense, or the change more acute, a local pneumonia results, and the cheesy deposit is met with. By thus restricting the term tubercle to a definite histological product, accompanying a well defined condition of body, and by giving a fresh name to processes which are different in their origin and causes, Dr. CLARK has thrown much light on a subject which has hitherto been very obscure. While the term phthisis may be kept as indicating a state, the chief symptom of which is a wasting away, it will be well if we recognize various forms of it, as different from each as the contracted gouty kidney from a contracted white one; then phthisis will be as genuine a name as Bright's disease, and the word tubercle will have a distinct annotation, and not, as heretofore, include a number of perfectly distinct pathological processes.

We will now take a view of the most approved and recent works on Tubercle, and we are done with this part of our subject. It is abundantly evident that a complication of effects produced by certain causes in the tissues of the body, with the causes of disease, has fallaciously guided the student of histology, instead of being led step by step up a gradual ascent from the study of physiology to that of pathological processes, and a spring is made over a wide gulf into a new and strange land. The truth is, that we pass by a scarcely perceptible transition from the phenomena of healthy nutrition to hypertrophy, inflammation, tubercle, tumors, fevers and other forms of perverted nutrition or disease. Hypertrophy may be what we call disease—this distinction being not one whit founded on instrinsic differences, but being an arbitrary distinction of our own, according to the usefulness or harmfulness as observed by us. Of the changes which take place; we see a blacksmith's arm enlarge as he works at his forge, and we see a thyroid or cervical gland enlarge from some cause which we cannot so clearly appreciate; we examine the condition of the tissues in both, and find in both a like state which we call hypertrophy, or hyperplasia. The elemental increase in both is identical; but we call the one healthy action, because we see why it occurs; while we denominate the other disease, because we cannot see its purpose in the economy, and therefore regard it as hurtful. From hypertrophies to tumors is but another step or shade of gradation. Between certain tumors, in their early stage, and products of inflammation, no distinction can be made; and we shall hereafter see how identical are tubercles with products of inflammation, on the one hand, and certain tumors on the other. Hence, from the merest hypertrophy, up to the most aberrant form of tumor, disease is nothing more than an unusual activity, or perversion of the very changes which are closely going on in the nutrition of the body, and which constitute what we understand by the life of the individual. Perhaps nowhere is the birth of tubercle better studied than in the pia mater of a person who has died of acute tubercular meningitis. A piece of this membrane spread out and examined under a moderately high power of the microscope, is found to be the seat of an extraordinary cellular tissue. The cells and nuclei which make this increase of proliferation, are seen partly scattered, profusely and at random through the membrane, giving it the well-known milky appearance that it offers to the naked eye, part clustered into little whitish or grayish knots, or particles, the individual tubercles. A close scrutiny of these latter will show that they are especially abundant along the smaller branches of the cerebral arteries, and seem here to take their origin in the cells of the connective tissue, which compose the adventitia, or outer coat of these blood-vessels. So marked is this the case, that although the tubercle-cells must be allowed to arise from the connective tissue-cells of the membrane generally, yet the adventitia appears to be certainly the part most actively concerned in their development. The tubercle-cells generally are rather smaller than a white blood-cell, have faintly granular contents, and are very brittle, so as very readily to rupture and set free their shining nucleus. Toward the border of the tubercle are seen cells larger than the above, often containing many nuclei, and manifestly containing connective tissue-cells in a state of active hyperplasia, or tubercular development. In and among the cluster of cells which compose the tubercle is a faint stroma, the original connective tissue, together with occasional blood-vessels not newly formed, for tubercle is non vascular, but belonging to the tissue in which the tubercle arises. Very much the same appearance is obtained from fine sections of liver or kidney, and from cerous membranes affected with tubercle, the connective tissue in all cases serving as the material for development.

In the lungs, wherever connective tissue is to be found, there may be tubercle present; but whether tubercle may also be developed out of the epithelium lining the alveoli, and grow inside the air-cells, is a point on which we find writers on this subject at issue. VIRCHOW denies that the little deposits found inside the air passages, and air cells, are tubercles. In his view, these are mere products of inflammation, or catarrh, which choke up the small bronchi, and on section so closely resemble tubercle proper, that they may be called spurious tubercle. He professes himself able to distinguish between these spurious tubercles, and the true tubercles of the connective tissue, a distinction which if it be real, must require unusual skill and assurance to make. VILLEMIN, on the contrary, mentions that these are positive intra-vesicular tubercles, which are no more the products of catarrh, than the extra-vesicular. Careful microscopic observation enables him to assert that the fine wall which partitions off the air-sacks from one another is not a homogeneous membrane, but encloses in its substance a cellular element, peculiar to itself. Under a high magnifying power, a beautiful network of capillaries is seen, covering this wall. Between meshes of this network he can detect a cellular element, occupying nearly the whole of each open space. There is, he asserts, no epithelial lining to the alveoli, but the tubercles are formed out of the above cells, and grow into the cavities of the air sacks, filling them up. LEBERT believes that pulmonary tubercles, or granules, as he calls them, take origin indifferently in the cells of the connective tissue, or in the epithelium of the alveoli, being both intra- and extra-alveolar. "If," he says, "you call the little particles outside the alveoli tubercles, you must call those likewise that are within, tubercles. They are all a part of the same disease, a result of the self-same irritant." BIZZOZERO, says, likewise the little intra-alveolar masses which accompany tubercle, are rarely tubercles, but spring out of the epithelium that lines the alveolus. In case of the sympathetic and ductless glands, VIRCHOW, though he alludes on more than one occasion, to the closeness of the link which binds the connective tissue with the lymphatic system, yet insists that it is in the connecting tissue frame work of these glands, and not in the gland-cells, that the tubercles originate. But VILLEMIN, looking at the connective tissue and lymphatics as parts of one great system, which he designates the lymphatic connective system, believes that tubercle may originate both in the cells and in the connective tissue of the lymphatic and ductless glands. Hence in tuberculous diseases of these organs, the microscope cannot of itself distinguish tubercle from simple hyperplasia, inasmuch, as elements identical with those already existing in health, are simply reproduced in excess. It is consequently a matter of extreme difficulty, nay, in some cases, impossible, to say whether a given change in one of these glands is simply hyperplasia, or tubercle. This question then must remain for the present an open one. That tubercle originates, mainly in the connective tissue, is the expressed opinion of all modern authors. That it may also find a means of development in epithelium and gland cells, is probable, but not yet clearly determined. Tubercle has the shortest life of any pathological product. *It is born but to die.* Hence its power of growth is exceedingly limited, and its hurtfulness when stationary but small. It is the expression of a general disease, and herein lies its deadliness. The same cause which produces it is circulating through the body, and may produce thousands similar to it, wherever connective tissue is present, and has, as VIRCHOW was the first to point out, manifestly malignant properties, spreading to the tissues in its neighborhood, and infecting distant organs by dissemination of its germs or juices. Thus, though there is a marked limit to the size of these masses, that are found by agglomeration of tubercles or to the number of tubercles, which may be scattered through the body, do we not see in these properties something that reminds us of the outbreak of an eruptive fever; and something too that recalls the behavior of cancer, subject to certain laws? An early tendency to die, is then the most characteristic and distinguishing mark of tubercle. Fatty degeneration beginning first at the centre of the knot, gradually spreads to the circumference, and gives it the well-known yellow color from whence the name yellow tubercle is derived, till little by little, the whole particle softens and breaks down into a cheesy granular debris.

We have therefore but to picture to ourselves a large cluster of tubercles, passing through this stage in the midst of some organ, as the lung, or kidney, in order to understand what wholesale devastation may be thus wrought, and how a cavity or vomica may be formed, in which are contained the remains of the softened tubercles, and of the broken-down tissues, involved by the tubercles in their destruction of the parts. On the other hand, if the part diseased be a free surface, as a mucous membrane, the disintegration and softening of the tuberculous mass will give rise to the formation of an ulcer, which differs from the ordinary process of ulceration, by the fact, that the base and sides of the ulcer are walled in by the remaining tubercular matter.

Foremost among the diseases, which either anatomically or clinically bear decided marks of relationship to tubercle, stands scrofula. This word scrofula seems to have been originally employed to designate a chronic swelling of the cervical lymphatic glands, in consequence of which the neck loses its contour and comes to bear some resemblance to the neck of a scrofa or pig. As this disease has been confounded with cancer, it becomes necessary to take some particular notice of it, and show wherein it differs from cancer.

With the enlargement of the glands of the neck, there are occasionally eruptions on the skin, or affections of the mucous membrane, and it was agreed to include them under the title of scrofula. After a while it followed that every glandular enlargement, no matter where situated, must be a manifestation of scrofula; certain internal glands, the bronchial and mesenteric, were now observed to be often in a state of hypertrophy and cheesy degeneration, while at the same time tubercles were present in the lungs or bowels. What more natural than to look at the ganglionic lesion and the tubercular products as the common effects of the same cause, scrofula? The thing seemed clear enough, especially when the French school detected the same specific cells in the cheesy scrofula of the glands, as had been found in the cheesy tubercle of the lungs and intestines. Thus the terms scrofulous and tubercular diathesis, were a convenient refuge for the destitute in all doubtful cases. Gradually, however, facts in contradistinction from this view, began to accumulate and to be noted down by independent observers. Foremost among those who combated the identity of the two diseases was GEUNER, whose masterly exposition of the subject in 1860 is well known to all medical readers. VIRCHOW, at the same time, completed GEUNER's clinical picture, by upsetting the doctrine of specific tubercle elements, and showing that the cheesy metamorphosis is peculiar, neither to scrofula, nor yet to tubercle, but is met with as a termination of simple inflammation of syphilitic tumors, or fatty buds, and sometimes of cancer. Both VIRCHOW and VILLEMIN look on scrofula and tubercle as distinct things. In the opinion of both pathologists, scrofula signifies a weak impressionable or vulnerable state of the body, in consequence of which irritants act upon the tissues with unusual severity and persistence. From this point of view, a simple glandular swelling following an ordinary irritant, is not scrofulous; but when a superficial irritation of the skin or mucous membrane persists, and is followed by swelling of the adjacent gland, which likewise persists, even after the primary irritation has been removed, we have a scrofulous person before us. Three great groups of glands may be affected with a scrofulous or scirrhous change: the cervical, bronchial, and the mesenteric.

Villemin is of opinion, that no anatomical distinction can be made between the scrofulous and tuberculous glands; that in both, there is simple hyperplasia, with a retrograde tendency. He thinks that it is in the exciting cause that the true distinction is to be found; scrofula being a mere tendency to chronic glandular enlargement, in consequence of slight external and local causes, or irritants in a child whose lymphatic system is unusually impressible, tubercle being on the contrary, due to a specific cause or virus, which provokes a more general outbreak. That which holds good for the glands, holds equally good for the bones, in which scrofulous disease is a mere chronic and very persistent inflammation, while tubercular disease is a specific lesion. Scrofula is, further, a disease almost peculiar to childhood; and is, therefore, intimately connected with the development of the body, and manifests its presence at a time when the lymphatic connective system is in the zenith of its functional activity. Tubercle, on the contrary, prevails between twenty and thirty, at an age when scrofula has ceased to disturb the body. Scrofula is not hereditary as a disease, but the exaggerated lymphatic vulnerability which favors its attack, is strongly hereditary.

Virchow goes still further, and finds a positive anatomical difference between the scrofulous and tubercular lesion in the diseased glands, as well as in the lungs. In the glands, scrofula is, in his view, a simple scirrhous hypertrophy of the glandular tissues generally, with a tendency to degeneration. Tubercle is a true neoplasm, forming in the connective tissues only, of the gland, and forming, save by extension, the proper glandular structure in the lung. Tubercle is in the connective tissue alone. Scrofula or scrofulous peripneumonia is a chronic inflammation of the bronchi and air-cells, which become stuffed up with their accumulated contents in a state of cheesy degeneration, and from cheesy deposits of larger or smaller size according to the extent of the inflammation. No one who has had much to do with children's diseases, can hesitate to accept the accuracy of this distinction. These cheesy particles in the lungs, accompanied by cheesy degeneration of the bronchial glands, without a trace of anything like a tubercle in any part of the body, are from time to time met with.

We cannot avoid, therefore, accepting a scrofulous pneumonia in youth, as distinct from tubercular disease; at the same time we agree with VILLEMIN that VIRCHOW and his followers ride their hobby a little too hard, in the case of the cheesy infiltrations which accompany real tubercular disease in adults. These, the Germans assert, are not softened tubercular masses, but are mere patches of chronic inflammation that accompany the tubercles, and have passed through the same cheesy metamorphoses, that tubercle itself undergoes. It is in vain you put before them their own assertion, that degenerate products of inflammation and degenerate tubercle are exactly similar in appearance, and that these patches are, therefore, at last as likely to be tubercles as inflammatory products, or that you point to the tubercles growing themselves around these patches, as if ready to form fresh ones. You are told, that what you see are not tubercles, but sections of small bronchi, filled with catarrhal accumulations.

LEBERT simplifies this matter most agreeably, by assuming that there is no such thing at all as tubercle, but that all these deposits are intra- and extra-alveolar, miliary and diffuse, gray and cheesy, nothing but inflammatory products. It is thus seen that tubercle and scrofula, though they have, anatomically considered, certain marked affinities with one another, are yet essentially distinct diseases.

It is in Glanders that VILLEMIN thinks he has found the closest marks of analogy with tubercle, not only in its anatomy, but also in its symptoms and causation. He seems to have been conducted from the study of glanders, directly to the inoculation of tubercle. The characteristic lesion of glanders, is a small tubercle, which is strewn either in the mucous membrane of the nasal passages, or in the lungs, or more rarely in the liver and spleen. At first a grayish white firm granulation, composed of cells and nuclei, apparently developed by hyperplasia of connective tissue. It soon tends to soften centrally, and forms ulcers in the mucous membrane and cavities in the lungs. Like miliary tubercle, it occurs isolated or in clusters. Together with this, little granulation streaks and bands of fibrous tissue, as well as patches of cheesy infiltration, are not unfrequently met with in the lungs of glandered horses. It is interesting to see that the same doubts have been raised concerning the real nature of these infiltrations in glanders as in tubercle. They are regarded by VILLEMIN as one form of glanders, just as in man they are one form of tubercle. As to which is the part primarily affected in glanders, the nasal membrane or the lungs, there is some difference of opinion.

VIRCHOW maintains, that the deposits in the lungs are always secondary, and by metastasis from the nasal membrane. PILLIPPE and BOULEY were convinced, by repeated post-mortem examinations, that the primary lesions are always in the viscera, more particularly in the lungs, and that the formations in the nasal membrane are invariably secondary. If, say they, a horse has the (Setage) discharge from the nose, he is already thoroughly glandered. It really matters very little, which part of the body is first affected. In either case, the analogy with a tubercular outbreak remains as strong as can be. The intestinal ulceration of tuberculosis, in which we see the counterpart of the nasal ulceration in glanders, is made often secondary to the pulmonary disease, but occasionally shows itself before any evidence of mischief can be detected in the lung. Again, glandular enlargement, of a severe and persistent kind, constitutes an important part of glanders as it does of tubercle.

The mode of invasion is also identical in the two diseases, now acute, destroying life in a few days, as by an overwhelming blood poison; now chronic, so as to last for years. Further, in the chronic form, the same occurrence of acute attacks complicating and adding to the chronic mischief, is observed in glanders as in tuberculosis. To read a description of chronic glanders, is *mutatis mutandis*, to read an account of chronic phthisis. It is therefore not surprising that DUPUY goes so far as to say that glanders is a tubercular disease in the horse. In speaking of the supposed cause of tubercle, we propose presently to follow out still further this remarkable thread of resemblance, in order to show the difference between tubercle and cancer cells. For the present it may suffice to say, that glanders is transmissible by inoculation, and contagious from horse to horse, and that it is also unmistakably communicable from horse to man.

"Can we hesitate to believe," says VILLEMIN, "that the parallel between tubercle and glanders must here find its completion? To conclude, glanders and tubercles are so closely akin, that they must be looked upon as nearly related species of the same disease. Syphilitic formations have in certain organs, more particularly the brain, so close a resemblance to tubercle, that the one may well be mistaken for the other. We see in both the same little cells and nuclei, forming apparently by prolification of the connective tissue, and heaped together in an intercellular substance; in both, too, a tendency to fatty degeneration, though less prominent in the syphilitic gummy tumor, than in the tubercular deposit. Syphilis must therefore take its place with glanders and tubercle, as one of a family" (doubtful). Who is there that has watched a case of tuberculosis from its outset, but must have been impressed with the remarkable similarity that it bears to the eruptive fevers, and more especially to typhoid fever? It is indeed doubtful whether in the first few days of the attack it be possible to make a diagnosis. In such cases, it is only as the disease makes progress, that special symptoms declare themselves, which enables a careful physician to recognize the presence of this most deadly of diseases.

The question as to how far tubercle and typhoid fever are antagonistic, is disputed. VILLEMIN is strongly in favor of such antagonism. He says, that "those who speak of fever as a frequent precursor of tubercle, have been led into error, by the great likeness of tuberculosis in its early stage to typhoid fever. What they have witnessed has been tuberculosis throughout, and not fever followed by tuberculosis." MURCHISON's authority is as strongly against any such antagonism. He says: "whether it be true or not, that persons laboring under phthisis are rarely attacked with pyogenic fever, an attack of pyogenic fever is often followed by tubercular deposits in the lungs." This is certain, however, that the postmortem appearances in the liver, spleen, kidneys and intestines of persons who have died of acute tuberculosis, have sometimes a most striking resemblance to the appearances in the same organs, after death from fever.

From cancer in its early stages, tubercle in its early stages is structurally undistinguishable. Both take origin from the connective tissue, and tubercle has been already shown to have malignant properties, which assimilate it to cancer. There are cases likewise of acute general cancerous eruption, which have some points of resemblance to acute tuberculosis. Cancer, however, has an innate vigor and power of growth and specific cells, which sufficiently characterizes it, as contrasted with the wiry degraded tubercle, and its kindred. Truly, tubercles are found in the lungs, the nasal membrane, as in consumption and glanders, and in scrofula, and occasionally in the pus secreted around cancerous ulcers. But all this does not prove that cancer is a tubercular disease. Although tubercle may be found in the adventetia, as well as cancer cells, yet it is distinguished by its being a tubercle, and cancer, by its peculiar and appropriate cells. Tubercle is only born to die. Cancer cells are persistent and aggressive, infiltrating and malignant.

TREATMENT OF CANCER

BY THE MOST EMINENT PHYSICIANS AND SURGEONS:

AS REPORTED IN THE PERIODICALS OF EUROPE AND AMERICA.

The first remedy I shall notice is the extract of hemlock, as given by Dr Osborn, of Dublin. He says: "the extract, imperfect as it is, has an effect in appeasing the pain in cancerous affections of the uterus, and that without exerting sensible narcotic powers, which almost excuses Stork for the error into which he fell, in proclaiming it a cure for cancer." I have applied it externally, and given it in such cases; sometimes without effect, and sometimes with remarkable alleviation of pain, after opium has failed, and I never observed any ill effects, except in one case, of a woman, laboring under scirrhous uterus. She obtained great relief from pain by it, but when the dose was increased to four grains three times a day, had headache, black motes in vision on sitting up, and saw two persons instead of one, all which disappeared when the remedy was discontinued. Let it be remembered, for internal uses, the extract made from the leaves should always be preferred, on account of their uniformity and perfect preservation. All the other preparations of this plant undergo decomposition, and consequently deterioration, and cannot be relied on for medical purposes.

M. Lisfranc used the knife in several cases of a very grave character, and with good success. Fortunately he removed all the tissue that contained cancer cells. But he says there are cases where the knife cannot be used, because the wound produced would be so large that the patient would not be able to bear it. He believes cancer may be produced by wounds or bruises, and in such cases the operation by extirpation is not likely to succeed. (Doubtful if cancer is ever produced in that way). He places more reliance on an operation for the removal of the diseased portions than most surgeons. He operated occasionally in the advanced stage and even where the neighboring lymphatic glands were enlarged. He did not always allow them to be an impediment to the operation, trusting to their dispersion by antiphlogistic treatment. Again, he says, a cancer may not be voluminous, but it may involve such important parts as to render its removal difficult or impossible by the means mentioned. The lessening tumors retire from the vicinity of the nerves, vessels, etc., which may have presented the obstacle to the operation. Some practitioners, observing the partial diminution of the tumor produced by these means, and ignorant of the principles upon which they have been recommended, reject the operation altogether, vainly believing they can dissipate the cancer wholly in the same manner, thus giving time for the disease to make fatal progress. He says cancer is liable to relapse. This is to be met by local depletion continued for some days prior to the operation. Relapses will be less likely to occur if half an inch of sound skin be included in the operation.

Dr. James Arnott's method of treating cancer by refrigeration is not without good effects. He uses pounded ice and chloride of sodium combined, applied to the part for five or six minutes. Its effects are immediate and completely anæsthetic. It supersedes the use of opium, morphia, or any other narcotic. His object is to reduce inflammation and allay pain and irritation. He is not without hope that it may by reducing the cancer cells finally destroy them, and thereby effect a cure. At all events, it relieves pain and saves the patient a great amount of suffering, for a period of time from ten to fifteen days; "nevertheless," he says, "the cancer may be making progress and advancing to its final fatal termination."

Why, oh, why should we stop when we have gained one point, and lose its good effect, when if the ideas by it suggested were carried out, our object might be attained! We shall see what can be done in the sequel.

Dr. FANCHOU, says the medical society of Paris, professes to cure cancer and several other intractable diseases. He presented two patients to the society, and requested the members to examine them. They did so, but strange to say, no two members could agree as to the real nature of the disease, nor would Dr. FANCHOU himself undertake to pronounce a diagnosis. He merely affirmed that they were cases of a *mauvaise nature*. However this may be, the Doctor proceeded to treat the patients after his own fashion, and again presented them to the society, after a lapse of some weeks; one of the patients had an open sore on the chin, which some said was cancer, others lupus. The second patient had beyond all doubt, cancer of the mamma, and was moreover, in a very bad state of health. Dr. FANCHOU brought back the first patient cured; as to the second one, the ulcer of the breast was nearly healed, the cancerous diathesis had disappeared, and the patient had become fat. These were incontestable facts. These results were obtained. The treatment of the lupus patient lasted three months and a half. It consisted of a substantial diet, and frequent exercise in the open air. These were aided by a great many remedial means. First, tincture of iodine, then Fowler's solution, and then arsenic and iodine, internally. The sore was occasionally washed with tar water, tincture of iodine, decoction of poppies, etc., and finally cauterized with nitrate of silver, creosote, and iodine; the cicatrix is now sound and healthy looking. The treatment of the second case was conducted on the same principles, but only gave rise to considerable amelioration. Even this, however, may be regarded as a progress, inasmuch as a majority of the members of the society had pronounced both cases absolutely incurable, when first submitted to them.

Dr. RIVALLIE treats cancer by solidified nitric acid. This is seized with long forceps, and placed upon the parts. After fifteen or twenty minutes, he removes it. An eschar four or five lines in thickness is thus obtained. There are some cases where he leaves the caustic twenty-four hours: When the cancer is encephaloid the pain is trifling, except where the skin intervenes. If the caustic is left a long time, there is no pain after the first three or four hours. With large diseased masses the application should be repeated every day. After the eschar has been carefully removed, the caustic is applied for a few minutes only, and the sore dressed with a solution of alum. Dr. Rivallie gives a few cases where this method of cauterization was used with success. (1849.)

M. Devay, of the Hotel Dieu, has long been engaged in investigating the therapeutical properties common in cancer. Being of opinion that Stork's experiments should be resumed, with the aid of the improved chemical knowledge of the present period, he finds the best preparation to be an extract and balsam, containing one per cent. of conium made from the leaves of the plant, gathered when at maturity, of full weight, and of ash gray color. In cancerous affections it exerts remarkable calming effects, and in some cases cures seem to have resulted from its employment, especially in the atrophied form of scirrhus. Its use is less satisfactory in soft and rapidly increasing tumors, but the progress of some of these has seemed to be retarded; in other cases it has diminished the size of secondary tumors, rendering the primary ones more amenable to surgical operations. As a means of assuaging suffering, whether used topically, or taken internally, it is invariably preferred by the patients to opium, and all other narcotics.

M. Manec, surgeon to the Saltpétriére, obtained 2000 francs, from the Academy of Science, for the care he has shown in investigating the action of Frère Como's arsenical paste in more than 150 cases of cancer, in some of which, he obtained unhoped-for results. His experience led him to these conclusions. 1st. That the arsenical paste penetrates the cancerous tissue, by a sort of special action, which is limited to it. This action is not simply escharotic, for beneath the superficial blackish layer, which the caustic has immediately disorganized, the subjacent morbid tissue seems struck with death, though it may retain its proper texture, and almost its ordinary appearance. Later the cancerous mass is separated by the eliminatory inflammation, which is set up around its limits. The same paste which extends its action more than six centimetres deep in a cancer of close texture, when applied to superficial gnawing ulcers, usually only destroys the morbid texture, however superficial it may be, and respects the sound parts. 2d. The absorption of arsenic is proportionate to the extent of surface to which it is applied, and as long as this does not exceed a two-franc piece in size, there is no danger from this source. A large surface should only be attacked by successive applications. 3d. That the arsenic which is absorbed, is chiefly eliminated by the kidneys, during a space of time, of not less than five, and not more than eight days, is amply demonstrated by PELOUZE. Thus if we allow nine or ten days to intervene between successive applications, all danger from absorption may be avoided. M. GOZZI also uses a caustic for the removal of cancer. He recommends the following: corrrosive sublimate, ℈j; caustic potash, ʒss; arsenic and ceruse, aa gr. vj; to be made into a paste with starch and white of eggs: while using this or other caustics emolient poultices, ointments, etc., should be avoided, as diminishing their effects, unless the irritation produced by these applications has been excessive. He objects to the plan of destroying the tumor, layer by layer, from the apex to the base, the latter becoming very indurated after these repeated applications, and offering great obstacles to the approximating surrounding granulations, and their cicatrization.

M. E. CAZENAVE speaks very highly of a caustic formed by pouring heated sulphuric acid on powdered saffron.

MR. MOORE recommends the chlorate of potash as a topical application in cancer. A solution in water, from one to two drachms to a pint of water, he found very useful in indolent ulcers, and phagadena, and in cleansing cancerous sores. He used it as an application to the mucous membrane of the nose, mouth and tongue, in cases of ozena and secondary ulceration. He suggested that the beneficial effects of the application were probably due to its seeking free oxygen, instancing its use in some forms of dysentery and other affections of the lower bowels. Mr. E. HAWKINS generally combined it with tincture of myrrh. The remedy is very useful in cases of cancer in removing the odor, independently of its effect on the sore itself.

DR. WEST, of St. Bartholomew's, almost invariably orders for his patients, who are the subjects of fibrous tumors of the uterus, a long course of one or the other of the preparations of iodine. The following is his prescription:

R. Potassii iodidi, gr. j.
Syrup. ferri iodidi, m. xx.
Aqua carui, f℥ss. M.
Ter die sumend.

He remarked at the time, that were the patient one in the higher ranks of life, it would be justifiable to send her to drink the Kreuznach waters. In common with DR. RIGLEY, and other physicians, DR. WEST entertains a high opinion of the value of the iodides in procuring the diminution of these tumors.

DR. WINN has used the Galium Aparine with good effect in cancerous sores. He says it grows abundantly in England (as it does also in this country). This plant is called in common language *Cleavers Bedstraw*. There are twelve varieties of this plant; it grows in dry and open grounds in the vicinity of Boston; stem erect; leaves in 8s grooved, entire, rough, linear; flowers densely peniculate; root long fibrous. Stem slender erect, 1 to 2 feet high, with short opposite leafy unequal branches; leaves deflected linear, with rolled edges; flowers numerous, small yellow, in small dense terminal panicules; flowers in June; the roots dye red. The flowers are used in England to curd milk.

The twelfth, or following variety, is the only one that possesses medicinal virtue. Its description is as follows: Stem weak, procumbent, reversely prickly; leaves 8 s, 7 s or 6 s, linear oblate, mucronate, rough on the midvane and margin; petal axillary 1 to 2; grows in wet thickets, in Canada, and Northern States to Indiana. Stem several feet long, leaning on other plants and closely clinging by their hooked prickles, to everything in their way; leaves twelve to twenty lines long, by two to three lines broad; flowers numerous, small white, etc., rather large, armed with hooked prickles; the root will dye red. When first Dr. WINN used the Galium, it was in the form of decoction, but finding this mode inconvenient he had an inspissated juice made. A teaspoonful of this juice is equal to half a pint of the decoction. In ordinary cases, a drachm taken three times a day, will be found sufficient, but in obstinate cases the dose must be doubled. Many patients have failed, no doubt, from not having given the remedy a sufficient trial. With regard to the efficacy of this plant in the treatment of cancer, Dr. WINN was solely indebted to the experience of Dr. BULLY, of the Bucks Hospital, who had forwarded to him an account of those cases of cancer in which the Galium had appeared to exercise a remarkable influence in arresting the disease. Large quantities of Galium are sold at Covent Garden annually, and it is chiefly used by the purchasers, as an external application in cancer. This fact corroborates Dr. BULLY's observation. M. REMILLY highly recommends the perchloride of iron to arrest the uterine hemorrage, which so frequently accompanies cancer of the neck of the uterus. He administers it by injections, in the strength of 15 of the perchloride, to 20 of water. He has used it in several cases with decided benefit. It not only corrects the fetor of the discharges, but prolongs the life of the patient. He does not, however, give a single case of cure from its use.

Dr. ROBERT JOHNS, in a paper read before the Surgical Society of Ireland, draws the following deductions from his investigations into the best modes of treating those diseases. 1st. The cancerous affections, when confined to the cervix uteri, are in many cases successfully treated by removal. 2d. That the only chance of preventing a return of the disease is to remove a healthy part of the cervix, from which it grows. 3d. The best and most expeditious operation, is amputation of the cervix, in a part free from disease. 4th. As hemorrhage is very likely to follow such an operation; a ligature ought to be thrown around the cervix, as high as possible, for 24 or 36 hours before amputation. 5th. The cauliflower excrescence is a disease to which this treatment is rarely applicable, as it rarely, if ever, extends beyond the neck of the uterus, and as it is one of the forms of cancer which is least liable to return, after the excision of the parts. 6th. Amputation of the cervix in hypertrophy, and such like affections of this part, which are curable by simple means, is not justifiable. 7th. Extreme prostration alone, or enlarged superficial inguinal glands, ought not to be a bar to the operation. 8th. As inflammation in many forms is likely to follow this operation, an appropriate preventive treatment ought to be adopted. 9th. The use of the actual cautery expedites cicatrization after the removal of the cervix. 10th. As amenorrhœa and dysmenorrhœa are likely to follow an extensive ulceration of the os and cervix uteri, and also when these parts have been removed, the uterine sound should be passed from time to time afterwards. 11th. With a view to correct the cancerous diathesis, the patient might be put under a course of treatment for some time previous, and subsequent to the operation. 12th. A particular form of vertigo is a frequent symptom, and an important diagnostic of uterine disease. 13th. As females suffer from uterine diseases which sometimes prove fatal, every means should be adopted to prevent their occurrence. 14th. In all cases of suppressions, or vaginal discharges, manual examination, at least, ought to be employed. We have given Dr. Johns' treatment for this formidable disease, and we see he has not succeeded in performing one cure by the use of the knife. He may in some cases have prolonged life for a short time, in a miserable state of existence.

Mr. Ellis says: the actual cautery of the ancients has given way to the more painful potential cautery of the moderns. M. Jobert was the first to recommend its application to the uterus in situ, in chronic disease of that organ. Mr. Ellis alludes to the researches of Mr. Marshall, who in his investigations on electric heat, in surgery, appears to have been the first who corroborated the views of the ancients respecting the remedial agency of the actual cautery He used only a heated wire, which necessarily acted on a small part or surface, and was consequently inefficient. The experiments of the author soon led him to adopt a better method, by which he was enabled to concentrate the heat evolved, over a considerable surface; an important element in electrical heat. The instrument employed was a good sized silver catheter, straightened out, with the end cut off, which formed the body of the instrument. It was then slit up at the upper end, and broached, so as to form a socket for the porcelain cauterizer, and also to allow the internal wires to pass out. Within the catheter is placed the two conducting wires, insulated, they being at one end connected with the wires of the battery, and at the other with a piece of platinum wire, which is coiled around the porcelain cauterizer. The battery employed is Grove's, with four or five cells. Of these, two are required to heat the porcelain to white heat, which degree is essential. From this simple contrivance the instrument derives its principal value. The heat being intense and permanent, when ready for use, it is entirely under the control of the surgeon; a matter of vast importance in its application. The patient to be operated upon, should be in the usual obstetrical position, and the batteries and wires concealed from her, so that she should not have any idea of the nature of the remedy. A good light and speculum are essential, the common glass one coated with gum-elastic. Neither the two-bladed, nor the conical glass forms are at all suitable. A full view of the os and cervix utori having been obtained, the os should be cleansed with a piece of cotton or wool, and when the cautery has become entirely heated it should be steadily introduced, and quenched in the diseased tissues, the duration of its application and the depth of its introduction depending upon the effect required. The eschar thus produced is marked with a whitish-yellow border, and the cervix often visibly contracts under the application of the cautery. The porcelain must be heated to whiteness, otherwise hemorrhage may occur, from the instrument dragging off a portion of mucous membrane, which invariably adheres to it. Under such circumstances, the surgeon should remember that the cases in which it is applicable are induration of the os, induration and ulceration of the cervix uteri.

Dr. Carmichael, of Dublin, says: "I have lately tried a new chemical combination of arsenic, mercury, and iodine, invented and recommended by Dr. Donovan, of this city. The instance in which it has been tried, is at present in the hospital, a case of lupus, engaging the nose and palate, of many years' standing, and which has destroyed the greater part of the nose. He has been using this remedy four weeks, and a most decisive change has taken place for the better; so much so, as to have surprised many of his confreres in the hospital. On the last consultation day, the ulcer of the nose was perfectly healed, while the patient himself is apparently much improved in constitution."

It is given in form of solution, and according to Dr. Donovan's communication in the 18th vol. of the Dublin *Medical Journal*, the proportions are as follows:

    R. "Water,   ʒj.
            Protoxide of Arsenic,   gr. ⅛.
            Protoxide of Mercury,   gr. ¼.
            Iodine converted into Hydriodic acid,   gr. 4-5." M.

The color of the solution is yellow, with a tinge of pale green. Its taste is slightly styptic. "It cannot be properly joined with tincture of opium, or with sulphate, muriate or acetate of morphia; for all these produce immediate and copious precipitates in it." Hence if opiates are to be used during the exhibition of this arsenico-mercurial liquor, they must be taken at different periods of the day. Tincture of Ginger produces no bad effect. Here follows Dr. Donovan's prescription from Mr. Carmichael, of Dublin:—

    R. Liquoris hydriodatis arsenici et hydrargyri, drachmas duas;
        Aquæ destillatæ, uncias tres cum semisse.
        Syrupi zingiberis, semiunciam. Misce.
    Divide in haustus quatuor; sumatur unus mane nocteque."

Thus we see "one-sixteenth of a grain of protoxide of arsenic, and one-fourth of a grain of protoxide of mercury would be taken in each dose, along with two-fifths of a grain of iodine, which being in the state of combined hydriodic acid, will be much diminished in energy of medical effect." This no doubt, is the proper dose to begin the exhibition of arsenic with: but it will very soon be necessary to increase it. The division into draughts is here necessary, first to insure accuracy of the dose, so essential in the case of this active medicine, and next to prevent injury to the ingredients by the use of a metallic spoon as a measure, the general way in which, unfortunately, the dose of medicine is determined. Thus Mr. Carmichael says, " I have been thus full in my account of this new remedy for diseases of a malignant character, to which class I am inclined to think lupus belongs, because I have every reason to believe, from the trial I have made of it, in this, as well as in some other cases, of a similar nature, that it will be found a most useful medicine."

Arsenic has been highly extolled as a cure for cancer; so much so, that it is in general introduced as the chief ingredient in all the secret plasters, or applications for this disease ; for instance, in the Plunket plaster, which is a favorite remedy, in all parts of Ireland. It occasions great pain and swelling of the surrounding parts, and causes deep sloughing of the cancerous mass, and is certainly often successful, when the cancer is of but little extent, draining out, as their phrase is, the roots of the cancer. When the carcinomatous substance is even less than the egg of a pigeon, Mr. Carmichael deems it not only a painful, but hazardous application.

The chloride of zinc has been within the last few years highly extolled as an escharotic for the cure of cancer. Mr. Cross and Dr. Ure have brought it into use. Recommended by such respectable authorities, Mr. Carmichael gave it a trial in several cases, from which he came to the conclusion respecting it, that though not so objectionable as the oxide of arsenic, on account of the poisonous nature of the latter, it excites just as much pain and inflammation without destroying the same extent of carcinomatous substance. He therefore did not feel inclined to try it again.

With respect to the class of narcotics, such as conium maculatum, belladonna, hyoscyamus, I shall briefly observe, that perhaps I have seen as many cases of cancer as any man in this country in private practice, and I have never seen any benefit derived from the use of these remedies. Further, their influence in mitigating pain, notwithstanding all that Baron Storck has said, with respect to their efficacy, I believe, that opium as a palliative and anodyne, is far superior to any of them.

Mr. Carmichael says, "I need now only briefly allude to the cases upon which alone I would permit an operation. But upon this head, perhaps, it is better to speak negatively. 1st. If the patient possesses that peculiar pale countenance which shows a predisposition to the disease; and particularly, if there is any reason to believe, on inquiry, that such predisposition is hereditary, I would decline to operate, no matter whether the axilla is free from disease or not, or that the tumor does not adhere to the pectoral muscles. 2d. I would decline an operation, if there were any general or physical signs that the lungs were not sound, or if there were reasons to suspect that they were tuberculated. 3d. I would also decline an operation, if the axilla or other lymphatic glands, in the neighborhood of the tumor, were hard and knotted, or if the breast was adherent to the parts underneath."

It follows from these objections to an operation, that the field for it in cancerous complaints is, in Mr. Carmichael's opinion, very limited; in fact, he would confine it to those cases where it either occurs directly from accident, if there be such cases, or in persons otherwise healthy, in whom the neighboring lymphatic glands are entirely free from disease. If these exceptions were attended to, we should not have those frequent acknowledgments of want of success, detailed by the most celebrated surgeons of Europe and America.

With regard to the case of lupus, given by Mr. Carmichael, with all due deference to his high character, we would suggest that seeing the good effects of remedies applied in effecting a cure, the case, in all probability, was not cancer, but syphilis. These remedies are truly anti-syphilitic, and cannot be relied on to cure cancer.

Of the treatment by Richardson's styptic colloid, I will give one case,—cancer of the mamma. Mrs. F. had first noticed a small hard tumor near the right nipple, four years ago. She was advised to have it removed, but she refused her consent. The tumor continued to enlarge, and in July last, ulceration commenced. The sore was five inches in length, extending from the nipple across the anterior border of the axilla; the remainder of the breast was closely adherent to the muscle. The axilla was filled with a mass of enlarged glands. The sore was sloughy, and very offensive. Continued lancinating pains deprived the patient of her sleep and appetite. She had lost flesh rapidly, and her face bore the appearance of cancerous cachexia in a high degree. She was at this time quite willing to undergo the operation, and begged Dr. Low to remove the breast. That, he told her, was impossible, as the disease had advanced too far. He could hold out to her no hope, except mitigation of pain, to make her last days more easy, and to remove the horrible effluvium, which made the patient a nuisance to herself and family. He painted the sore over with colloid, and gave her a bottle to take home with her. He also gave her a morphia pill, to be taken at night. She came again, the following week, when he was surprised to find the sore had become quite clean, and was healthily granulating; every trace of smell having disappeared, and, more, the pain had also gone. She had enjoyed comfortable nights. Her appetite had improved, and with hope in her countenance she looked years younger than she had done the week before. She felt certain she would get well, but on this point he could give her but little encouragement, so extensive was the amount of disease. However she continued to improve. The sore which was constantly painted with the styptic, gradually filled up. No pain was experienced, except when she run short of the dressing. The large mass of axillary glands diminished pari passu with the sore, so that now they can scarcely be felt. The sore itself is not larger than a four-penny piece, at the end of a year, and would, he thought, have been quite healed, but for the difficulty in inducing her to rest her arm, which, owing to her having a large family of small children, she has not been able to do. This is a very interesting case, and certainly shows the benefits of the colloid dressing.

Dr. Atlee, of Philadelphia, treats both cancroid and carcinoma of the uterus by the long-continued use of arsenic internally, in small doses, combined with the local use of a very strong solution of iodine in glycerine,—a drachm of iodine and a drachm of iodide of potassium dissolved in two drachms of glycerine. This is applied with a brush or cotton, two or three times a week, all over the cervix uteri and to any part of the growth that can be seen. He gives one case of undoubted carcinoma uteri, where the patient's husband, being a medical man, fully confirmed all Dr. ATLEE said of the remarkable improvement which had followed this treatment in her case.

Dr. MADDEN, Surgeon to the Cancer Hospital, London, uses a paste made of arsenic and gum acacia: two drachms of the former to one of the latter. This paste he applies to any part of the body where cancer may be situated, except the mouth or nose; but if the cancer exceeds four square inches, he applies it only on one-fourth the surface at a time. When a cancer exceeds this limit, he knows of no means for its removal but the knife. Again, he says, "It must not be supposed, because I so strongly recommend the arsenical mucilage, that I would discard the use of the knife altogether. In some cases it is our only hope." He applies the paste to any part of the body except inside the mouth and nose,—parts, in fact, where the use of the curative agent would be dangerous. He does not recommend its use when the disease is deeply seated; but for any cancer on or near the surface this mucilage is the least painful, and the most certain remedy of which he knows. For the last seventeen years he has tried every known caustic, and now firmly believes this is *the best*. After the cancer is killed, he poultices it for some days, until it falls out. If granulations spring up too luxuriantly, he treats them with styptics, and keeps them down until the sore heals.

Dr. J. R. WOLF, Surgeon to the Aberdeen Infirmary, and WEEDEN COOK, have almost ignored the knife in the treatment of cancer, and use styptics and escharotics; for instance, the sulphate of copper, chloride of iron, and the chloride of zinc, which they find more reliable than the knife. Still they do not lay the knife aside in all cases. The best surgeons, both in Europe and America, are seeking for a more reliable remedy than the knife for the treatment of cancer.

DR. BROADBENT gives the history of an interesting case of cancer of the breast. In 1864, by his advice, the cancer was removed by MR. W. COULSON. The disease returned, and was again removed in 1865. In May, 1866, the tumor was growing more rapidly than ever, near the cicatrices of the former operations. It was decided that no further removal was advisable; and unless something could be done, a miserable fate was before the patient. The hypodermic syringe is now in the hands of every physician, and it seemed to him that some fluid might be injected into the tumor, which would so far alter its nature and so modify its nutrition that its growth might be retarded, or at least arrested. After considering the various substances which presented themselves to his notice, he selected acetic acid, for the following reasons: 1st. This acid does not coagulate albumen, and might, therefore, be expected to diffuse itself through the tumor, and the effects would not be located at the point injected. 2d. If it entered the circulation, it could do no harm in any way. 3d. Acetic acid rapidly dissolves and modifies the nuclei of cells, on the microscopic slide, and might be expected to do the same when the cells were in situ. 4th. It had been applied with advantage to common ulcerations.

On May 18th the first injection of the acid was given. The tumor was of the size of a small egg, and a patch of skin about the size of a shilling had become adherent to it. The needle was introduced through sound skin, an inch or more from the part involved in the disease, and passing to the centre of the mass. About thirty minims of diluted acid (one part of acid to two of water) were injected. It gave little or no pain. The next morning a bullæ containing dark bloody fluid was found to occupy the patch of adherent skin.

May 23d.—This portion of the skin dry and hard; the adjacent part of the tumor not so hard. Again injected the acid.

The patient residing in the country, was not again seen till June 7th, when the piece of skin mentioned was found detached from the surrounding sound skin, and a probe could be passed in all directions to a distance of three-quarters of an inch or more between the tumor and sound structure, a little discharge issued from the fissure mentioned. Injected the acid on this date, and again on the 9th. The acid used being rather stronger, it gave a little pain, and swelling and tension of the parts around followed.

On June 11th, and a few days afterward, there was a free discharge of fluid and solid portions, with relief to the swelling. No fetor whatever attended this discharge, which afterward diminished greatly.

June 26th. On external examination, the tumor was found to be much smaller, and, in passing a probe into the opening, it entered a large cavity extending on all sides. Part of the walls seemed free from malignant structure; but at several points a deposit of cancerous matter remained. On attempting to inject the acid, it was found too thin to retain the fluid, which either entered the tissues, and gave great pain, or made its way into the cavity. The cavity was stuffed with lint, saturated with acetic acid, and the case left in the care of a family medical attendant, who was to inject the acid as he saw necessary.

July 13th. No impression made on the remaining disease, which had, in the opinion of the medical attendant, extended somewhat. Carbolic acid was tried for a few days, as an application, but discontinued, and the cavity dressed daily with strong acetic acid, and injections used daily. This energetic treatment gave much pain, and excited inflammation all around.

August 4th. There had been considerable hemorrhage, which had been arrested by susquichloride of iron. The result, however, was removal of the entire remains of the malignant disease, and when last seen, a healthy granulating surface was visible at every point. Three other cases are related by the author, and some conclusions drawn as to the cases in which this mode of treatment is applicable. In the use of any new remedy, we should give the negative effect as well as the positive.

Our object being to give every one a fair statement and

trial, so that every advantage may be had that is attainable, we therefore give what Dr. Charles H. Moore has said, on the use of acetic acid. He makes the following remarks upon the treatment of primary tumors:—" There is one question before all others, relating to the treatment of primary cancers, which calls for attention, and which I therefore refer to in the present communication: it is the injection of such tumors with acetic acid. It is not yet possible to foresee the extension of which that method of treating cancer is capable. I have myself much hope from the employment of it, and I am always satisfied with its effects in secondary tumors, but it has not yet been made applicable to the primary. From the first promulgation of this treatment by Dr. Broadbent, I have never used it, nor advised it in any case for which the ordinary operation by the knife was adapted. It did not to me appear right to try a remedy so little tested and the adequacy of which for relieving all conditions of the disease was uncertain. Already I have known that both disappointment and damage have resulted from an experimental use of the acetic acid, in cases of primary cancer of the breast. It is the more incumbent on me to say this, as by announcing the destruction of cancerous matter in the interior of a lymphatic gland with this acid, and the absolute dispersion of small recurrent subcutaneous cancerous tumors, by the same means, I may unwittingly have encouraged others to what I cannot but think a misapplication of the remedy. So ready a method of disposing of cancer is not yet attained. Its unseen diffusion beyond the apparent limits of the tumor, is too certain a fact to justify confidence in injections for the removal of it. Neither is it yet to be expected, of a remedy so slow in its action, and the management of which is yet so far from being perfected, that it should all at once supercede the more certain operation. The condition of primary tumors appears to me to make them particularly unfit for this treatment; for whilst they are growing, and may become large, the acid can only be thrown into them in small quantities, and at intervals. If used in large quantities, it produces suppuration, sloughing and disastrous action, of a remedy in a primary cancer. In any quantity it produces swelling with consequent uncertainty as to the area over which the effect of the acid has been secured, and delay in pursuing the treatment. Meantime the tumors continue to grow, in the parts concealed by the swelling. I cannot think this to be the right treatment of a primary cancer." Such objections may appear to relate no less to a secondary recurrent or advanced, than to a primary tumor, but the circumstances are in fact quite different. The treatment of the latter disease is avowedly undertaken with less prospect of advantage than that of the former. In those advanced cases, the acid can achieve the reduction always, and sometimes the removal of the morbid mass. It is only in such cases, where established methods of treatment are unsatisfactory, that those which make greater promise ought to be proved.

Dr. Dieu, of the Hôpital des Invalides, has used acetic acid with the happiest effects, on those tumors and ulcers of the skin which are classified under the heads of cancroid, nolimetangeres, and cutaneous epithelioma; one case of which was of eleven years' standing. The ulceration of the lower eye-lid and angle of the eye was as large as a franc-piece, yet after ten applications of the concentrated acid, during one month only, cicatrization was complete. Another case in an old soldier, aged 77, a tumor of twelve years' standing, as large as a filbert, with an indurated base, was completely removed after two injections, and four applications of the concentrated acid. Three other cases were treated with similarly happy results.

Dr. Richardson has used his styptic colloid in open cancer, with good effect. He applies it all over the surface, with a soft brush, and lets it remain till some indications appear of matter or other fluid oozing from the edges. He then removes it, by applying tepid water until it becomes soft, and reapplies it, continuing to do so until the sore is healed.

Dr. Routh says the use of bromine was first suggested to him by Wynn Williams. The two cases are recorded at the Smithsonian Hospital. In the first the patient was thin, pale, and haggard, losing blood continually. There was a mass of fungoid epithelial growths, taking their origin from the os uteri, and about the size of an egg. The actual cautery was used to check the bleeding, and after the slough had come away, a solution of bromine, five minims to fifty of wine, was used. Pieces of lint were saturated and applied to the surface, and kept in situ by pledgets of lint. After forty-eight hours it was removed, and the part dressed with a pledget of lint dipped in tepid water. Occasionally warm douches were applied. In about a week a slough came away and left a large granulating healthy surface. Tannin with glycerine was then applied, and used daily. The iodide of arsenic and conium were given internally. After a period of twelve weeks she was fat, hearty, and of good color. As she occasionally lost a drop of blood, he (Dr. Routh) carefully examined the internal surface of the uterus, and found about a quarter of its lining membrane affected with epithelioma. The patient left the hospital for some weeks, and on being re-admitted, a piece of wood about the size of the uterine cavity was prepared, and covered with cotton. The upper part was dipped into a saturated solution of carbonate of soda; the lower part into the bromine solution. It was passed up, and left within the uterus. Two or three further applications of bromine with glycerine were necessary, and the patient left the hospital with a remarkably healthy uterus.

The second case presented a large carcinomatous mass about the size of an orange attached to the os uteri, which appeared to be a large cauliflower excrescence, breaking down readily and bleeding at the slightest touch. The mass was removed by the wire écraseur, and four days after, the spirituous solution of bromine was applied. She took internally, the iodide of arsenic, and conium, and was treated in the same way as the first case, and left the hospital fat and hearty.

TUBERCLE.

In the opinion of some writers, tubercle, or consumption, scrofula, and cancer, are only varieties of the same disease. We beg leave to differ from those who hold this opinion; and, in order to show why we do so, we will offer the following remarks, to show the distinctive points of difference. And, first, of tubercle. Every formation which is pathologically different from any other formation, manifests this difference, even in its determinate microscopical structure. The invariable elements of tubercle are molecular granules, hyaline, ligamentous, or connecting matter, and the tubercle cells peculiar to tubercle. They are very small, irregular in shape, containing in their substance no nuclei, but molecular granules.

In water, ether, and weak acids, they undergo almost no change. In the concentrated alkalies, as liquor ammoniæ, caustics, and aqua potassæ, they are completely dissolved. The dimensions of tubercle cells are liable to manifold variations, which depend neither on difference in organs nor on age. They are discovered earliest in yellow, crude tubercles. Tubercular corpuscles are cells left remaining at a low stage of development. The opinion that tubercular substance is a modification of purulent matter, is most decidedly contradicted by microscopical examination. Tubercular corpuscles are distinguished from imperfect pus globules by the special shape of the latter, and by their larger diameter; and from perfect pus globules by the nuclei appearing in the latter; and, lastly, distinctly from carcinoma cells by the latter being from two to four times as large, and consisting of cell-membrane, of a larger distinct nucleus, and often even of nucleoli. In the softening process in tubercle, the connecting matter is fluid; the corpuscles are rounded off; their close juxtaposition ceases; they become expanded, and seem to be large. This, however, is incipient disintegration. The purulent matter surrounding softened tubercle never has its origin in the tubercle itself, but uniformly in the parts immediately surrounding it. The microscope, in doubtful cases, is necessary in distinguishing whether we have to do with softened tubercle or with thickened tubercular matter, or with a mixture of the two. Purulent matter appears to cause the disintegration of tubercular corpuscles, and therefore to render them not cognizable in their individuality; while irregularity in outline and compressed mutual aggregation indicate the first stage in the development of tuberclecells; and the mutual softening, expansion, and rounding off, denote the second stage, while the third stage consists in solution. The minute globules are separated into a semi-fluid granular mass, and lose their individuality. The induration and calcareous transformation of tubercle is one mode of natural cure; the peculiar tubercular elements are wasted, and are partly absorbed. In their place are substituted small mineral granules, and sometimes crystals of cholesterine. The process of calcareous transformation is usually attended by the deposition of coloring matter, according to the chemical analysis of BOUDET. The chief elements found in these calcareous fragments are chloride of sodium and sulphate of soda. Calcareous salts are only found in small amounts. In tubercles, we specify as elements, not constantly but occasionally present, melanosis as the most frequent admixture, and, next to this, fat fibres. Dark, olive-colored globules and crystals, we find only accidentally mingled with tubercles, but by no means belonging to their substance; they are products of inflammation, as exudation and suppuration, and the elements of epithelium fragments, in multiplied forms. The seat of tubercle in the lungs is usually the elastic tissue of these organs. These bodies are, however, also found scattered in the pulmonic vessels and in the capillary and minute bronchial tubes. The pulmonic tissue around tubercles may be sound, but for the most part it is in a state of congestion or of inflammation. The later morbid process is either globular, or diffused over the greater part of an entire globe. The purulent matter found surrounding tubercles, is often not the consequence of any hepatization, but proceeds from the mucous membrane of the small bronchial tubes, partly destroyed, and opening into the tissue of the lungs. Peripneumonia around tubercles possesses nothing specific. We find in it the same elements of exudation, aggregation to globules, fat vesicles, purulent corpuscles, etc., as in ordinary pneumonia. Tubercular corpuscles are not usually found among exudation products. Occasionally there is found around tubercles a peculiar form of chronic inflammation, with yellowish hepatization, and with increased consistence in texture. The pulmonic vessels, the small bronchial tubes, and the pulmonic tissues, are filled partly with fibrinous coagula, with new fibrous formations, partly with aggregate and purulent globules. Amidst chronic hepatization with few vessels we find vascular, acute, lobular pneumonia. The degree of consistence of acutely and chronically inflamed lungs, depends on the amount of fibrous matter, fluid blastema, and globules contained. Much fibrinous matter, with little blastema and few globules, produces induration. Much fibrous matter, with little blastema and few globules, produces softening. An equal proportion of these elements produces a middle degree of consistency, the lungs rendered compact in consequence of compression from above, by pleuritic effusion if present. Nowhere is any inflammatory pneumonia. The gray semi-transparent granulations of the pulmonic tissue contains from the beginning tubercle-cells, and are therefore a genuine form of tubercle.

Their coloring and transparency is produced partly by the close juxtaposition of the tubercular corpuscles, in consequence of uninjured pulmonary fibres; partly by the presence of a large proportion of connecting matter. The gray granulation is not always the termination of the formation of yellow tubercles; the latter is often developed primarily as such. The vascular network found around the gray granulations, is neither a sign of inflammation nor consequences of new formations, but rather the effects of compression of many capillaries by the tubercular deposition, and the consequent over-distension of the residual capillary vessels, reduced in amount. The notion that gray granulation may be a consequence of inflammation, is contradicted by correct research. The process of ulceration is altogether different from that of suppuration. Thus we find on the mucous membrane of the bronchi suppuration without the formation of ulcers, and, on the mucous membrane of the intestines, ulcers without suppuration. The last cause of suppuration is effected by inflammation, in consequence of parasitic deposits, occasionally from obliteration of a certain order of capillary vessels, produced by causes hitherto unknown. The tubercular ulcer of the lungs is, physiologically, not different from the tubercular ulcer of the bowels, and of the skin.

In tuberculosis, a general ulcerous diathesis is formed, particularly in organs in which tubercles very rarely take place. This fact is well established by Louis. The internal fluid layer of the contents of ulcerous cavities of the lungs contains the following elements: 1st, Tubercular substance, seldom entire; globules, for the most part in the process of expansion, or in that of granular dissolution; 2d, pus globules, occasionally in smaller amount; 3d, fused globules; 4th, aggregation of globules; 5th, purulent pus; 6th, blood globules; 7th, pulmonary fibres; 8th, black coloring matter, or melanotic deposits; 9th, epithelial fragments; 10th, occasionally, crystals; 11th, adipose tissue. Among this thickish fluid are found, for the most part, false membranes, consisting of fibrinous matter, inclosing purulent elements. Among the false membranes covering the diseased pulmonary tissue, is found a true purulent membrane, consisting of vessels and a fibrous basement membrane, inclosing minute globules. This is usually partly destroyed by fresh tubercular eruptions taking place amidst it. This membrane is the result of a healing effort of nature, in order to isolate the ulcerated pulmonic tissue, and thereby to favor its cicatrization. Between the purulent membrane and the pulmonic tissue is often found a layer of newly formed fibrous tissue. Around the ulcerated cavities deposits of new crude tubercles for the most part take place. The healing of cavities is accomplished—1st, by the isolating faculty of the purulent membrane, and the contracting of the circumference of the cavities; 2d, by fibrous matter, which fills the cavities, unites with their walls, and thus forms a fibrous cicatrix; 3d, by deposition of saline matter in the cavities, and formation of fibrous tissue around them. There are no such formations as muscular corpuscles. That which has been described as such, are merely pus corpuscles, secreted by diseased mucous membranes. Proofs of purulent matter are therefore useless. In the expectoration of phthisical persons is found the following elements: 1st, Mucus; 2d, pus corpuscles, uniformly in large quantity. These are occasionally found in a shriveled state, and may then readily occasion error; 3d, Epithelial fragments in different forms; 4th, granular matter in great quantities, consisting, probably, of dissolved tubercular corpuscles; 5th, small yellow pellicles, portions of false membrane; 6th, pulmonary fibres; 7th, adipose vesicles; 8th, blood globules, occasionally with fibrous coagula; 9th, occasionally, aggregation globules, small infusoria, or vibriones. Proper tubercle-cells are not usually found in the expectoration of phthisical persons. There is, therefore, no uniform or certain indication by which to distinguish the expectoration of pulmonary consumption from that of many other diseases of the lungs. Pulmonic fibres in the expectoration denote the presence of ulcerated cavity. Their occurrence is, nevertheless, rather exceptional than constant.

The greatest part of the expectoration of phthisical persons proceeds not from cavities, but is secreted by the bronchi. The abundant secretion of mucus and purulent matter, often taking place from the bronchi, in pulmonary consumption is one of the ways which nature strikes out in order to prevent greater disturbance in the circulation, which would be a necessary consequence of the imperviousness of one set of capillaries, and the over-distension of the rest. One part of the dissolved tubercles of the ulcerated cavities, is mingled with the expectoration, another part is absorbed. The law enunciated by Louis, that after the age of fifteen years, the lungs contain tubercles if they are found in other organs is, on the whole, correct. It may nevertheless be modified in this manner: that provided in any other organ a very considerable deposition of tubercles takes place; for instance in the liver, the kidneys, the peritoneum: then the lungs will often contain a few tubercles. In childhood, tubercles in the cerebral membranes, in the glandular system, and in the peritoneum are more frequent than in adults. The thickening of the pleura, from pulmonary tubercles, has its cause, not in inflammation, but in increased nutrition; since it becomes more vascular, in so far as it receives a portion of the pulmonary blood, and therefore becomes a supplemental organ to the circulation through the lungs, and at the same time, by agglutination with the walls of the chest, enlarges anastomotic communication with the large circulation. From embryological and pathological researches it results, that neither around tubercles, nor in the false membranes of the pleura, are new vessels formed, independent of the general circulation. Uniformly new vessels are formed in diseases in the centrifugal direction from the large circulation. The apparent conversion of false membranes into cartilaginous substance, depends only on the fibrousness of compression, without secretion of proper cartilaginous elements. In like manner the so-named ossification of false membranes, consists only in amorphous deposition of mineral formations. The three chief forms of glandular tubercles, are those of the superficial or subcutaneous glands, the bronchial glands, and the messenteric glands. The last possesses least tendency to softening. Tubercular matter is in the glands throughout, the same as in other organs. The existence of a sensible, demonstrable, scrofulous matter, we cannot admit. That which is regarded as such, is either the consequence of ordinary inflammation or suppuration; under the influence, no doubt, of a dyscrasial element, yet without peculiar matter, or of inflammation, or suppuration, attended by tubercular deposition. Tuberculosis is, in the osseous system, a much rarer disease than observers are disposed to allow. Very frequently concrete purulent matter, and tuberculous matter, are in the same case confounded with each other. In doubtful cases the microscope alone can solve the difficulty. From actually scrofulous diseases, which for the most part become known by inflammation products, and suppuration, tubercular diseases are to be distinguished on the one side, and on the other, chronic idiopathic inflammation of the eyes, of the skin, of the glands, of the bones, and of the joints.

The last category, often in children, is attended with scrofulosis. In short, accurate diagnosis and definite conceptions of scrofulosis, is a desideratum, daily more urgently demanded in modern medical practice. The gray granulations in the cerebral membranes, particularly the pia mater, show distinctly tubercular corpuscles between the fibres of the serous membrane. They further appear in the brain, frequently with yellow miliary tubercles, and with tubercular inflammation also along with large tubercles. In the liver, tubercles appear occasionally in very considerable masses, and even with true cavities. These cases are rarely confounded with cancer. So far as the medullary fungus of the liver, when converted into softening, and disintegration has been examined, an appearance very similar to tubercular deposition has been observed.

Fatty deposition in the liver, and fatty organization in the heart, occasionally takes place in phthisis; consequently there is a tendency to the deposition of fat in the internal organs, while it undergoes wasting in most parts, particularly in the external, forms a pathological character of tubercular and other diseases. The kidneys may be also almost, or quite, filled with tubercular deposition. In cases of this class, few tubercles are found in the lungs. In tubercles of the peritoneum, we find among the tubercular corpuscles many fibres of serous membrane. Tubercles of the peritoneum are little disposed to softening. They are for the most part accompanied by considerable deposits of coloring matter. Tuberculosis of the peritoneum occasionally causes perforation of the intestines, which is for the most part fatal; yet in instances of great rarity, by means of the formation of an artificial anus, a preservation of life is still permitted. The consistence of crude intestinal tubercles, is normally less firm than that of other organs. In the tubercular ulcers of the intestines, no purulent matter is found. The microscopical elements of tubercular intestinal ulcers are, besides the disintegrated tubercle-cells, cylindrical epithelial fragments, granular disintegrated mucous membrane, and remains of fibres and fasciculi of the muscular coat. The new epithelial cells should not be confounded with pus corpuscles. In the morbid intestinal mucous membrane of phthisical persons, we find occasionally, polypiform, melanotic and tubercular excrescences. Tubercles are found in very rare cases, deposited between the walls of the arteries; a part most favorable to the excretion of tubercles from the blood, and one equally strong in opposition to the synonymousness of cancer-cells and tubercles in the pericardium and in the heart. Tubercles are found there, and in that case very extensive adhesions and vascular anastomosis between the branches of the coronary artery, and those of the surface of the lungs; a remarkable communication between the vessels of the large and the small circulation.

Tubercles in the thoracic organs, as well as those in the abdominal organs may open passages externally, and thus form pulmonary as well as intestinal fistulas. Tubercle and carcinoma do not naturally exclude each other, and therefore do not avert each other's progress. Both morbid processes may at the same time run their several stages of development in the same individual, and yet the evidences of the existence of the one, is separate and distinct from the other, and in no case is their analogy so strong as to satisfy, for a moment, that they are only varieties of the same disease, as some writers have attempted to prove.

My only apology for entering so fully into this discussion, is to show, when we come to describe cancer and its cells, that tubercle and cancer are separate and distinct diseases, and that the remedies adapted to the cure of the one, are not, and never can be used successfully in the treatment of the other. Tubercle is only born to die; for as soon as it is perfected it dies, and is ready to be discharged. But cancer-cells are not so. They are tenacious of life, persistent, infiltrating, and do not die easily. Nothing but their complete removal by death, or otherwise, will arrest their destructive and dire influence.

In the above remarks on tubercle, we have quoted largely from LEBERT and others, to whom credit is hereby given.

CANCER DEFINED.

In order to come to a correct knowledge of cancer, its formation, and nature, we must first endeavor to ascertain where and how cells are formed in animal bodies; then ascertain the character and effects of those cells. The communication of E. Montgomery to the Royal Society, will throw much light on this subject. The organic cells, chiefly those of various cancerous tumors, are seen on the addition of water, to expand to several times their original size, and at last to vanish altogether into the surrounding medium. The nucleus does not always participate in this change, but sometimes remains unaltered, whilst the outer constituents of the cells are undergoing this process of expansion.

This curious phenomenon of extreme dilatation is intelligible only on the supposition that the spherical bodies in question are in reality globules of a uniformly viscid material, which by imbibition swell out, until at last their viscosity is overcome by the increasing liquefaction. In embryonic tissue, and in various tumors, single nuclei are seen, each surrounded by a shred of granular matter. On the addition of water, there will emerge from one of the margins of the granular mass, a clear globule, which continues to grow till it becomes a full sphere, which ultimately detaches itself, and is carried away by the currents.

At other times no such separate globule would be emitted, but the entire glandular shred would itself gradually assume the spherical shape, and ultimately encompass the nucleus, constituting with the same the most perfect typical cell. Not only single nuclei are

FORMATION OF CELLS.

found surrounded by a shred, but also clusters of two, four, and more, are seen similarly enclosed, by a proportionally large granular mass.

Under these circumstances, sometimes by the addition of water, the whole granular mass of such a cluster, becomes transformed into a large sphere, containing two, four, or more nuclei. The resulting body will be to all appearance identical with shapes well known by the name of mother-cells. In all those cases the granular shred must have partly consisted of a viscid material, which, on imbibition, naturally assumed the spherical shape. MONTGOMERY then proceeded to his experimental verification. In all the observations cited by him, the existence of a viscid imbibing material was proved with almost conclusive evidence; a material, which is capable of forming globules of a definite size, and while in the living organism, actually forming such globular shapes, the nature of which, has, perhaps, been entirely mistaken. After a long search, the substance known under the name *myeline* was found to be the desired material. When to myeline in its dry amorphous state, water is added, slender tubes are seen to shoot forth, from all the free margins. These are sometimes wonderfully like new tubes in appearance. They are most flexible and plastic. From their curious tendency to shoot forth in a rectilinear direction, it may be inferred that a crystalization force must be at work. To counteract this tendency and to oblige the substance to crystalize into globules, it must be intimately mixed with white of egg. The result will be most satisfactory. If you mix with clear serum, you will have the same results. Hence you may see how a cancer-cell is formed. Instead of tubes, splendid clear globules; layer after layer is formed, resembling closely those of the crystalline lens found under similar conditions. Here is actually found a viscid substance, which on imbibition, formed globules of a definite size. The remaining task is comparatively an easy one. By mixing myeline with blood serum, globules are obtained, showing the most lively molecular motion. Thus, says MONTGOMERY, cells being merely the physical results of chemical changes, they can no longer afford a last retreat to those specific forces called vital.

Physiology must be something more than the study of the functions of a variety of organic units. "Any pathology will gain new hopes in considering, that it is not really condemned to be the interpreter of the many abnormities to which the mysterious life of myriads of microscopical individuals seem to be liable." (*Medical Press and Circular*.) The stellate corpuscles called plasma-cells, and connective tissue corpuscles are not permanent elements properly belonging to fibrillary connective tissue—but really transitory elements, proper to elastic tissue, and in which the existence of a cavity can by no means be demonstrated. The primitive fibres of fibrillary tissue, called connective, do not possess a central canal—of which there are no means of demonstration, nor are the elastic fibres channelled.

Dr. Belajeff investigates this subject as one of particular interest to physiologists. He says, "The lymphatics of the penis of man presents a true network of separate closed tubes, the walls of which have an epithelial lining. The chief of these capillary networks extends beneath the mucous body of *Malpighi*, where it forms a layer of numerous ramifications and anastamoses. More deeply, the lymphatic branches become larger and rarer. Vertical sections show their very large gaping openings, with an adherent internal epithelial wall. Near the surface the tubes are, as I have said, thinner, smaller, and forming sometimes meshes in the furrows of the skin, at other times little closed prolongations amongst the papillæ. The diameter of the largest lymphatics is, one to two millimeters; that of the smallest 0.08 mm. Although this variation of size is considerable, it is yet always less than that of blood-vessels. As a proper characteristic of the lymphatics, I would mention dilatation is to be observed, either about the middle of the channels, or toward their confluence. They may be circular, or total, or they may be uni-lateral in their lymphatic capillaries. They do not indicate the presence of valves, as in the lymphatic trunks, they only bear reference to a simple dilatation.

"The single epithelial layer of the capillary tubes is of oval, polygonal, fusiform or indented cells. The longitudinal axis of the cells corresponds to that of the vessels. The nearer a capillary tube is to a trunk, the more are its cells complicated, and their form elongated. The terminal meshes, on the contrary, have tolerably large and rounded cells. The flattening of the walls of a transparent tube gives to the cells, under the microscope, a multiform appearance, for the dark lines of the borders of the cells cross reciprocally, and those which belong to one of the walls, modify the normal appearance of the cells of the wall beneath it. The average length of the cells is .06 to .04 mm., and their width is from 0.008 to 0.020 mm. There is greater abundance of lymphatic capillaries throughout the glans, and the balano-prepucial furrows than in the penis and the prepuce. The epidermis layer has no lymphatics; they are beneath the mucus body of Malpighi, whilst below them, the network of blood capillaries is very abundant, even to the summit of the capillaries. The lymphatic trunks of the penis have two important characters, without reckoning their considerable size,—the frequency of varicosities, and the composition of their walls, which are formed of two or three layers above the internal epithelial layer cells, and annular fibres accompanied by fine cells, said to be of the connective tissue, present themselves. The elastic, and perhaps, muscular fibres, are in the elongated spaces which exist between the rows of transverse or annular cells. Their direction is sometimes rectilinear, sometimes zigzag.

"If we take a thin section from the centre of a sore, or tumor, which we suspect to be cancerous, and examine it with a power of 250 diameters linear, it will be found to consist of a meshwork of fibrous tissue, forming waved bands arranged here and there in circles, varying in size. Some of them will be very large, one-fifth of a millimeter in diameter, enclosing other circles, each of which will be surrounded by several filliments of the fibrous tissue. Some of these circles contain numerous nucleated corpuscles crowded together, mixed with granules; others contain only a few, or nothing but granules. Here and there, will be seen several compound granular corpuscles.

On adding acetic acid to the whole structure, it becomes more transparent. Many of the fibrous filliments become invisible, and such as remain are studded here and there with elongated nuclei. The walls of the corpuscles are partially dissolved and rendered very transparent, where their nuclei are unaffected. The surface of the tumor, for the most part, contains nucleated cells, compound granular corpuscles, and numerous granules. The nucleated cells are of a round or oval form, varying in size from the 1-100th to the 1-50th, or less, of a millimeter in diameter. Some will contain one nucleus, others, two, of an oval form varying in the longest diameter, from 1-100th to 1-75th of a mm. Some of these nuclei will contain one nucleolus, and others two. On the addition of acetic acid, the cell wall will be rendered more transparent, the nuclei will be unaffected and appear in consequence more distinct." This presents a good example of the true cancerous structure, and the cells described are the true cancer cells.

Let us see how far these observations will enable us to determine more satisfactorily than before, how to recognize cancer, before and after death. To do this, we must look at the matter physiologically and pathologically, as related to cancer. Then we shall be the better able to adopt the most appropriate and successful treatment. To fully recognize a cancerous growth, we must consider the common subjective and objective symptoms, such as lancinating pains, unequal surface, hardness, an elastic feel, ulcerative affections of surrounding parts, and constitutional cachexia. All these symptoms do not always attend cancer, being occasionally absent in tumors, which are undoubtedly cancerous. Therefore a critically microscopical examination is necessary, in order to decide this momentous question. This, with such evidence as may be afforded by the symptoms and progress of the case, is absolutely necessary, before a correct diagnosis can be given.

What then are the microscopical characters of cancerous growths? If a tumor presents, among other elements, certain cells of definite shape, or aspect, and at the same time is situated in parts which render it impossible that these cells can be epithelial, epidermic, or cartil.

age cells, that tumor must be cancerous. The cancer-cells are of variable shape,—round, oval, caudate, heart or spindle-shaped, and of variable size, from 1-100th to 1-10th of a millimeter; nucleated, single, double, triple, more generally nucleated; but colorless or with melanic deposit, increasing commonly on an endogenous plan, and thus often presenting the phenomenon of parent-cells, containing two or three generations of younger cells; occasionally increasing by division of the nucleus, but never by division of the cell wall, or by a splitting up of segments of the mother-cell. The cell walls are dissolved by acetic acid, the nuclei corrugated, and their margins thickened. Cancer-cells never pass into fibres; they become caudate, or throw out pointed prolongations, but the transformation proceeds no further. They are contained in great abundance in the milky juice, which can be squeezed from most cancerous tumors, and they are surrounded by a greater or less quantity of viscous fluid. In these cells may be found epithelioma, embrional and cartilage substances. They cannot be confounded with any other species of cells. They are infiltrating and persistent, and if one cell be left, in an attempt to remove them, it will generate and multiply others, until the cancer breaks out again, in all its malignancy, and destroys the patient.

But a few more evidences on this subject. Parent cells containing within them sub-cells, having darker nuclei, and these in turn bright nucleoli, are strongly characteristic of cancer—though such cells may possibly be absent in fibroid cancer, and hematoid may in some of its parts be devoid of them. The shapelessly caudate cell is significant of cancer. A tumor may present to the naked eye, the characters of hematoid, be the seat of hemorrhage, affect the communicating lymphatic glands, run, in all respects, the course of cancer, and nevertheless contain no cells, but such as are undistinguishable, in our present state of knowledge, from exudation cells. Nay, more, while a primary malignant tumor contains these cells alone, the lymphatic glands, secondarily affected, may contain compound nucleated cells, and spherical and shapelessly caudate cells. The true fusiform cell is an adventitious formation, when it occurs in cancer, and therefore has no special diagnostic signification.

Although we are of opinion that there is no one constant microscopic character which will serve as an infallible mark of cancer in every case, yet we are disposed to attribute the highest diagnostic value to cancer cells, as made manifest by the microscope. They will be present in a great majority of cancer growths, let the class of cancer be what it may; yet we cannot admit that their absence in some rare cases negatives the existence of cancer.

We agree with DR. BENNETT relative to the use of the microscope as a means of diagnosing during life, before the tumor has been meddled with, and when doubt exists as to its real nature. M. KUSS, of Strasburg, has proposed to extract a portion of a doubtful tumor with an exploring needle, and to examine microscopically the portion thus cut off. In the case of open sores, a microscopic investigation of the discharge will often clear up all doubts as to the cancerous nature of the disease, by detecting at once the characteristic cells.

It will be found, in a great majority of cases of open cancer, that, by squeezing it, a milky, albuminous juice will appear. This is a single evidence, other than that by the microscope, that will decide the character of the sore. Occasionally, however, no fluid can be expressed, as in dry fibroid cancer. DR. BENNETT gives a case of cancer of the mouth, where no fluid could be expressed; and yet, on scraping the surface, numerous cancer cells could be detected by the microscope. In some cases, enchondroma gives by pressure a milky juice.

Another and very important character, by which cancer may be distinguished from all other growths, is its property of infiltrating the tissues among which it lies. It extends its meshes in all directions, and in some cases, when it is situated in the liver, produces a transformation of the adjoining tissues into carcinomatous matter. Nevertheless, cancer is sometimes circumscribed, and in a few cases may even be encysted, and even when it is infiltrated it cannot be recognized during life. But, logically, this is a characteristic of great importance, and we are safe in attributing great importance to this peculiarity; and although infiltration may occasionally be found belonging to other tumors, yet its value is not weakened when applied to cancerous growths. I am disposed to agree with Dr. Walshe, when he intimates that infiltration, with common exudation matter, has been mistaken for fibrous infiltration. It has been asserted that it is a peculiarity of cancerous growths to return after extirpation; and this fact strongly affirms the infiltrating, persistent, and vital character of cancerous formations.

It is by no means an easy task to give the physiological and pathological history of cancer. Those gentlemen who have charge of large cancer hospitals in Europe,—men who are the first in the profession,—have better opportunities to set this matter in a clear light than those whose opportunities are comparatively limited, and especially in a place where every man who attempts to treat cancer is posted as a humbug or charlatan. The lazy indifference and ignorance that causes such remarks and prompts to such conduct, only deserves the profound contempt of all who wish to be benefactors of the human race. Every effort to throw light on a disease that has been an opprobrium medicorum, should be hailed with satisfaction; and, although perfection may not as yet be attained, yet every step toward that object is worthy of our regard. As light is here and there shed upon the subject, it only opens the vista through which others may see, and further truths may be developed, and science advanced, till ultimately a remedy may be found that will eradicate this dire disease. We return to the investigation.

The first step in a cancerous formation, is an exudation of the peculiar blastema which in our present knowledge can be distinguished only from common nutritive blastema, by the difference in the resulting organic forms. Such a blastema, as a general rule, exudes completely from the vessels, is seldom arrested in the thickness of their walls, and very rarely if ever, remains within the vessels themselves. This cancerous blastema thus derived from the blood, necessarily implies that the blood itself is in an abnormal state, when it can furnish so unhealthy a substitute for the ordinary secretive and nutritive blastema, and this inference is corroborated by almost all the facts we can bring to bear on this point, and which all tend to the conclusion that a cancerous tumor under all circumstances is but the local evidence of general vitiation of the system. Subsequently, the cancerous deposit may remain quiescent; the cancerous cachexia which produced it having passed away; or it may advance with greater or less rapidity according to the intensity of this cancerous diathesis. If it increases, it is destructive, or malignant, in proportion to the degree of the infiltrating power possessed by the cancerous tumor in the particular individual. Finally, the local disease itself reacts on the constitution, and impresses a deeper and more advanced influence on the general system. The great majority of facts with which we are acquainted, lead to the conclusion, that the filaments, cells, and fluids, which together make up the tissues which we call cancerous, originate in a coagulated exudation. This is poured out exactly in the same manner as all other forms of exudation, viz., by enlargement of the capillaries, their repletion with blood, and the transudation through their coats of the transparent liquor sanguinis, which coagulates outside the vessels constituting an exudation more or less solid. The exudation thus produced which we call cancerous exudation must differ from the exuded matter that is called inflammation or tubercle; but in what this difference consists we are not so thoroughly satisfied. Although the present state of pathology does not warrant our stating positively wherein exudation differs, we cannot avoid seeing that its occurrence in various individuals produces very different results. Thus, if an exudation be found in a healthy individual, it produces a series of phenomena, which we call inflammatory; if in a scrofulous individual, another series of changes, called scrofulous, or tubercular; whilst in another person, the result may be cancerous growth. Every kind of reasoning must lead us to the conclusion, that these different changes and effects depend, not upon the vascular system, which is the mere apparatus for the production of exudation; not upon the nervous system, which leads to the necessary arrangement of that apparatus, for the purpose, and not on the texture, which is the seat of the exudation, as that varies, whilst the cancerous formation is the same; but in the inherent composition itself, or constitution of the exudation. But we must not forget, that the blood itself is dependent for its constitution on the results of the primary digestion, in the alimentary canal, on the one hand, and the secondary digestion in the tissues on the other.

Dr. BENNETT remarks that "a comparison between the characteristics of the cancerous, and what may be termed simple or inflammatory, and the tubercular exudation, may lead to some useful hints respecting the significance of each. Let us suppose that an individual in perfect health has from any cause received an injury which produces inflammation. There is an exudation poured out, which is plus the normal exudation required for the repair of the tissues. The further changes which occur in this, vary infinitely, according to the part, the degree of irritation, the rapidity or slowness with which the blastema is poured forth; but one or other of what may be termed the healthy or normal changes occur. The exudation becomes the medium of adhesions, or is organized into false membrane, or accumulates in the interstices of organs, assuming greater or less resemblance to them, and impeding more or less their functions. Or it breaks down, more or less rapidly, into evanescent cell forms, which are discharged at once from the free surface, or are eliminated from the body through the medium of further changes and different emunctories. The composition of this blastema no doubt varies infinitely within the healthy limits; and these variations are individually not dependent, for the most part, on any local condition of the part in which the blastema is exuded, but are attributable to the general condition of the organism, to the purity and excellence of the blood, which is the indication of the due performance of all the varied metamorphoses of the frame."

If, instead of this healthy constitution, the individual be tuberculous or cancerous, then in certain organs the exudation emitted from any cause does not pass through this series of normal changes, but assumes one or the other of the forms peculiar to the diathesis. Thus, in young subjects, and in persons of peculiar and defective assimilating powers, the exudation is below the healthy standard. It attains imperfectly to any cell form, and is liable to rapid disintegration, and complete loss of vitality. This is the tubercular exudation; and, although it is poured out in its characteristic form in only certain organs and positions, yet we see enough to convince us that a blastema poured out in other situations, as in accidental breaches of continuity, although it may not possess the tuberculous character, is yet often individually different, in some way, from the reparative blastema which would be effused in a perfectly healthy condition of the system. In an older person, in whom the assimilating functions are also deranged, although in a different manner, the blastema poured out may be entirely dissimilar to the tuberculous blastema. It may have a tendency to the production of very highly developed and reproductive cell forms, which grow with great energy and rapidity, and retrograde and disintegrate slowly and imperfectly. Between this diathesis and the former there is, therefore, almost an antagonism. They can hardly exist in the same person, and the presence of one may be said almost to exclude the other. There the hypothesis we have given above agrees with the facts in the case; for it is well known that tubercle and cancer appear at different periods in life, and affect different organs; while, when recent, they are rarely found associated. May not tubercle be connected with some derangement of the primary, and cancer with the functions of secondary, digestion? For cancer commences at that period of life when the excretory functions begin to lose their activity.

Háve we not said enough on this part of our subject? It is hoped that some light has been shed here which will lead others to pursue the subject, and make further developments, throw new light, and bring out new truths. There is no limit to our science. The microscope will be to us what the mariner's compass is to the navigator. By a judicious use of it, we shall be led into the hidden arcana of nature, and physiology and pathology will find a rich reward from its developments. We have men in the profession, of mind and indomitable energy. They are the great benefactors of the human race, and their reward is sure, though it may be delayed. Thousands will rise up and call them blessed. The anxious thought, the racked brain, the sleepless hours, and days and weeks of vigilance and research, will meet its reward. The watchword is, "No failure."

We have been often asked this question:—What is the cause of cancer? Once for all, we do not know. Indeed, we do not know the remote cause of anything. It requires a Deity to know these things. We only know the proximate cause of any disease; and if we have correct apprehensions of the proximate cause of disease, and have a correct knowledge of our Materia Medica and the modus operandi of the medicines we use, and prescribe with a correct physiological and pathological knowledge, we have a fair opportunity to arrest the progress of disease and restore our patient to a state of health. But to return to the question. Habit, diet, constitution, predisposition, and location may have more to do with the generation and development of cancer than they may at first be supposed to have. The medical world is alive to these inquiries, and research will develop facts but little thought of heretofore. Science is progressing, and facts are stubborn things. The developments of physiology, materia medica, and chemistry will finally make these subjects plain, when preconceived opinions, and old threadbare theories, will yield to the development of true knowledge, and many a disease that has hitherto been an opprobrium medicorum will yield to the true light of science, and our profession will rise higher and higher in the scale of science, and the whole world will be blessed thereby.

Dr. HAVILAND has given us some very interesting facts relating to the distribution of cancer, as we have them recorded in the *Medical and Surgical Reporter*, of Philadelphia. He says he has found, by an examination of the Registrar-General's Health Statistics, that cancer is a disease which is among the most baffling with which medical men have to deal. It is of comparatively rare occurrence in the northwest of England. Now Wales and northwest England belong geologically to the oldest formations, the silurian and the carboniferous; while the other great districts free from cancer include the chalk-hills of Southern England and the öolitic formations of Gloucestershire, Northamptonshire, and Oxfordshire. Stronger still, in fact a good deal stronger, is the evidence furnished by the arrangement of those portions of England which, also free from cancer, connect the two great districts we have mentioned, which he has colored on the map blue, and says this is a sign of their exemption. These blue strips extend across the red sandstone of Warwickshire; the red sandstone and lias of Leicestershire; the red sandstone and carboniferous rocks of Derbyshire, and the sandstone heights of Worcestershire and Staffordshire. The two blue patches and their connecting links—on which take place but a small number of the forty-two thousand deaths from cancer which occur in the kingdom of Great Britain year by year—cover those parts of the country which are of the most ancient geological formations. Turning to the regions where cancer is most fatal, we find them to belong to the tertiary or most recent formations; and we find, also, that for the most part they surround the great rivers after they have reached their full size and approach the sea. The Thames cancer-field is said to come to its bounds where the London clay stops, the clay being of the lower eocene, a late formation. Other cancer fields are in the alluvial valley of the Humber and in the alluvium of Norfolk, Suffolk, Lincolnshire, and East Yorkshire. Now the distribution of consumption is almost exactly the reverse of that of cancer. It flourishes on the old formations, and gets its smallest number of victims in the lower and more sheltered, though damper haunts of cancer. That these haunts are lower and more sheltered may, to be sure, be the reason of their comparative exemption from phthisis; and the fact of their geological structure may bear only secondarily on their character as seats of disease.

If these facts show anything that should govern our opinions as to the locations where cancer is most frequent, it may lead to an inquiry into the products of these geological regions, and throw light upon the ingesta and digestion of these products, and lead to something in the regulation of the primary ingesta that will have a satisfactory effect upon the fluids in the secondary digestion.

CLASSIFICATION OF CANCER.

We have clearly distinguished cancer from all other classes of ulcer, and shown that this disease is distinctly defined by its peculiar cells, and these cells are distinct from tubercles—that tubercles are only born to die. As soon as they are formed, they cease to grow; they lose their vitality, and are dissolved, and ready to be thrown off. When they soften and dissolve, they form cavities, and by suppuration are discharged, and others form and follow the same course. This process goes on step by step, till hectic fever is set up, and the patient dies. Moreover tubercles are not persistent and infiltrating, but soon soften, and mix with the secretions of pus from the walls of the cavities they make, and are discharged. Cancer cells are not so. Let their origin be what it may, they are found in the adventitia, or on the skin immediately under the cuticle. Whether they are first formed here, or deeper seated in some of the faciæ, or in the lymphatic vessels, or in the fibrous tissue, or in the periosteum, as in some cases of hematosa, they in every case show their undisputed character by their peculiar cells. These cells have been clearly described in the preceding part. When these cells burst, they discharge an ichorous and corrosive fluid, which is capable of destroying the healthy tissues, that lie in contact with them; the lymphatic vessels, which are distributed to every part of the surface, and the extremities of the whole vascular and organic system become the subjects of their destruction, although the lymphatic and absorbent vessels resist their influence for a given time. Yet, by the force of their malignancy they ultimately cause those vessels to take up the cancerous fluid, secreted by their cells, and perhaps many cells with it, and pass them to other localities in the course of the lymphatics. Hence we often see the glands in the axillæ filled with cancer cells, conveyed from a cancerous mamma, and even when the cancer is located in a part where the lymphatics are not centered in the neighborhood, we find that the cancerous influence has been distributed among the muscular fibres, for some distance from the margin of the sore, and even to penetrate the bones. Hence the everted edges and thickened integuments in and near the cancer. So minute and infiltrating are these cells, that the finest absorbents may take them up, or fibre, tissue, ligament or bone may be penetrated by them, and yet their influence is so resisted by them that they make slow progress in their work of infiltration, so that years may pass, in some cases, before they become destructive to the neighboring tissues; and in other cases their progress is more rapid, so that within a few months they destroy a large surface. In all cases it will be found, that the infiltration has pervaded a third or more of the part than the surface of the sore presents, or the tumor will measure. Where cancer-cells are found, be they many or few, they give a specific character to the disease. If we see a child, a lad, a man of ordinary size, or a giant, we pronounce them human beings, regardless of their size, and to belong to the genus Homo. So with cancer. If it is an epithelioma, first found under the cuticle, or a fungoid or spongy sore, or a fibroid or hard sore, and deep-seated, or a hematoid, taking its origin deep, and perhaps from the periosteum or some deep adventitia, swelling up and bleeding freely on the least abrasion—if we find the characteristic cells in them, be they few or many, that disease is cancer, just as certain as the various sizes of the human species all belong to the same class of animal life. All these classes of cancer, if let alone, will terminate in the death of the subject.

In order to cure any of these varieties of cancer, we must remove every cell, and when the sore is healed, the patient is cured, and not without it. If you use the knife, you may get them all, but experience has abundantly proved that this is a rare occurrence.

Drs. Bennet and Walshe—who have done so much for the profession, and proved themselves great benefactors of the human race, by so fully and clearly describing cancers, and designating them by their cells, that no man free from prejudice or preconceived opinions can doubt the truth of what they say—have plainly shown that the knife is not a certain remedy for cancer. Also, the great American surgeon, S. D. Gross, although he says the knife is the only remedy, also says, if you leave one cell, the disease will return. If that gigantic mind was only free from the knife on this subject, for a while, and would turn his attention to those remedies that are calculated to remove those malignant growths by following them to their utmost extent, he might crown his already world-wide reputation with an additional wreath of fame. But the medical world is alive to the subject, and as we gain a little more knowledge, let us hand it out, and where one stops, let another take it up, and as our materia medica, chemistry, and pharmacy are not limited, truth after truth is being developed by the research of men of mind. Let us hope the day is not far distant when a remedy will be found for this dire disease, and instead of thousands dying yearly of cancer, such an occurrence will be as rare, as for a man to die of any of the exanthematous diseases.

It occurred when I first commenced the practice of medicine, more than fifty years ago, that I was called on to treat cancer. I knew of no remedy but the knife; and in the space of twenty-five years I cut out, as I thought, a large number of cancers. I am fully persuaded I did not cure ten per cent. of them. I became sick of the work, and set my mind to find some other remedy for this horrible disease; and whether I have succeeded or not, one thing I know, I have not for the last twenty years lost ten per cent. of my patients, including those cases that were too far advanced, when application was made to me, for any treatment to succeed. I have cured hundreds, that have remained well for years, there being no sign of a return of the disease. I do not claim to have discovered a remedy that is not known to the profession, but to have taken those remedies that have been in use, and so compounded them as to triple or quadruple their powers, and then applied them in a different manner from that in which they have ordinarily been used.

The great point to be attained in treating cancer is *to remove every cell* and diseased fibre. Having done this, we are sure of a cure. No internal remedies will reach them, and no external remedies that fail to effect this will accomplish a cure. This can only be done, as far as our knowledge now goes, when the cancer is so situated that the remedies can be applied to it. Certainly, constitutional treatment is often indispensable; but when the cancer is located on some internal organ, and the remedy cannot be applied to it, I confess I know no remedy that will reach the case. But we are rather anticipating our subject. We should have a clearer understanding of the physiological location, structure, functions, and classification of the parent gland cells, as found in the human system, and then we shall be able to see more clearly what lies before us, and what we have to do, and why we should do it. The simplest form of a gland is a parent cell, each one performing glandular functions; hence, every parent cell may be called a parent gland cell. So long as the parent gland cell organizes normal products in the normal quantity, no such state as disease of any kind can occur. As soon, however, as those normal conditions are disturbed, the physiological balance is deranged, and pathological states arise, at first functional; but if long continued, they tend to organic changes, all more or less constantly exposed to conditions which have a tendency to disturb the physiological functions. But little control can be exerted by constitutional remedies over the cell elements and the filamentous tissues formed by the metamorphosis of these cells, after they have been organized and hence escaped from under the influence of the parent gland cells. It is upon the parent gland cells that all healthy, as well as diseased, impressions are made, and it is to these cells that we look for the causes of diseased states; and to them we must address our remedies, if we expect to remove the cause of pathological derangements. Generally speaking, the physician is not called upon to treat cancer till the primary or formative stage is past, and the first step that should be taken is lost; and he then has to treat the resulting disturbance, which, to all intents and purposes, is the specific disease.

DR. SALISBURY says, "There are six types of parent gland cells in the human body," four of which it is important to understand, in order to a proper understanding of the classification of cancer. These cells organize all the tissues, products, and elements of the human body. 1st. The parent epithelial gland cell. This is the seat of epithelial cancer. These cells organize all the products of epithelial tissue. They are extra-vascular, and rest upon the outside of the so-called basement membrane. These cells in the milk glands organize and eliminate the milk; in the sudorific glands, sweat; in the sebaceous follicles, sebaceous matter; in ceruminous glands, cerumen; in the plain surfaces, pigmentary matter, and products to be eliminated; in the hair follicles, they organize hair; in the salivary glands, saliva, and the resulting products; in the mucous follicles, mucous; in the pepsin glands, pepsin; in the epithelial portions of the liver, bile; in the pancreas, the pancreatic secretion. For the mulchefying fats in the glands of BRUNER and LEIBERKUHN, the products mainly for digesting farinaceous food; those covering the villi feed upon and transmit materials to the more highly animalized gland cells inside the basement membrane. Wherever we meet with epithelioma, it will always be found located, in the first instance, in, or attached to, this structure of cell formation. 2d. In the parent fibrin gland cells, which line the red and white blood apparatus, organize all the products of blood and striated voluntary muscles. It feeds upon albuminous products, and organizes them into fibrin. This is the seat of fibroid cancer; hence the appropriate classifications. 3d. In the parent involuntary muscular fibre gland cells, which organize the involuntary muscles. This is the seat of fungoid cancer, and gives its classification. 4th. In the parent connective tissue gland cells, which are situated inside the basement membrane, and outside the parent cell layer that lines the red and white blood apparatus. These are the cells that organize all those products and tissues, the distinguishing proximate principles of which are either gelatine, chondrin, or ostein. From these cells emanate connective tissue, cartilage, and bone. From this location arises hematoid cancer (fungus hematoides); hence this classification.

This is as far as our subject is connected with the cell formation of cancer. Perhaps there is not one case in a thousand that does not commence as a local disease, and if properly treated in time may be cured. Many cases, no doubt—some from repugnance to the idea of having a cancer—are let alone till the absorbents have taken up the cells and cell matter, and they have been deposited in some neighboring gland or internal and vital organ. Then all treatment will be unavailing. Many others have been deterred by their friends from having them treated, by telling them the less they have done to it the longer they will live; and after it is too late for any remedy to reach the case they apply for aid. Then the ignorant on the subject will say, " Those who pretend to cure cancers are impostors," when in fact they are the cause of the delay that resulted in the patient's death, by keeping him from the use of the proper remedies, while there was a chance of being cured.

When I became satisfied that extirpation by the knife was a doubtful and unreliable remedy, my mind was turned to escharotics. I used a number of them. They all consisted of caustics of such description as were then known; but they generally failed to effect a cure. Lunar caustic would only kill the surface, and throw off a thin crust, leaving the principal part of the disease untouched and deep-seated. I was at that time ignorant of the true nature of cancer. I did not understand that a cancer was a disease *sui generis*, that no other disease was like it, and it like no other. Nor did the profession at that time, as far as I knew, know any better. It was taught that almost any inveterate sore might finally turn to a cancer. Consequently, if the surface of the sore could be changed and a healthy granulating process set up, cicatrization would ensue and a cure be effected. Hence the knife, as the quickest and shortest remedy, was invariably resorted to. But since, the question has been settled that cancer is a specific disease, and can be certainly known by its specific cells;

where these are found, be they many or few, the case is cancer, and where they are not found the case is not cancer, nor ever will be. Cancer cells never appear as a secondary development; they are always primary, persistent, durable, and infiltrating. These facts being fixed in the mind, we at once look for a remedy that will not only act upon the surface of the cancerous tumor or open sore, as the case may be, but one that will reach the utmost extent of the infiltrated and surrounding integuments, so that every portion of the disease may not only be reached but eradicated. As before stated, if one cell be left, the cure is not effected; for such is the vitality of these cells, if one be left it will recuperate, multiply, and break out with all its power of malignancy, and in many cases with renewed force.

The question arises here, Shall we use the same remedies in all cases? The answer is, Where the same remedies can be applied so as to reach every part, they should be applied; for cancer is cancer, whether it belong to the first, second, third, or fourth class. It is, however, true that we cannot apply the same remedy, in the same form, to every case. Nor is it necessary always to apply the same remedy, even in cases of a like character. But be the remedy what it may, it must be one that will reach every cell, and not only stun them but literally kill them, and remove them from their locality. These remarks apply to cancer when it is so situated that it can be gotten at, for when it is located in an internal organ I know of no remedy that will reach it. But perhaps not one in a thousand are located internally. They almost always make their appearance on the surface, either in the mamma or on the face, nose, neck, and rarely on the extremities, or in the tongue or gums. From these points they may spread, and finally pervade large surfaces, and by being absorbed, the neighboring glands become diseased, and even the cells and fluids may be carried by the circulating vessels, and deposited in some internal organ. But rarely, if ever, do we find them first developed in the internal organs, except the stomach, bowels, and uterus; and these organs may, in a physiological sense, be considered as external, for the same mucous membrane that covers the external surface, extends through the mouth, down the esophagus, and lines the stomach and bowels, and also lines the vagina and uterus.

Having said thus much we proceed to describe epithelioma. This is the mildest form of cancer. It shows all the true cancer cells, hence it is cancer. It generally makes its appearance first on the cheek, nose, forehead, or temples; but it does appear on other parts of the body, on the hands, arms, mamma and uterus. It often commences by forming a thin dry scale; this after a time falls off, leaving a small red spot, for a short time, till another scale forms on the same place. It sometimes itches a little; finally after scale after scale has formed and fallen off, it leaves a little dry sore, and a little matter is secreted under the scab: it enlarges by slow degrees, till a permanent sore is established, the edges of which very often become everted, and the sore secretes a creamy pus, which has a peculiar smell, characteristic of cancer, and which differs from the odor of all other secretions. If you remove this scab, you will find a sore under it, that presents numerous granulations. Take a small portion of these granulations, and place them under the microscope, and you will find them mostly composed of cancer cells. With a powerful glass, you may see all around the sore, under the cuticle, and in the true skin, that the vessels are enlarged and red, or of a purplish color. To a greater distance from the sore, you may not be able to detect them with the naked eye, but your glass will reveal them, and where there are two or three in the same neighborhood the communicating vessels may be traced. Let this sore alone, and it will spread and deepen. It may be months or years, but sooner or later the edges become everted, hard, and almost entirely composed of cancer cells. Ere long it begins to show its corrosive character, and with astonishing rapidity destroys the skin and muscles, till all the parts in the neighborhood are destroyed. All these months or years of dormancy, it has been making its way under the skin, with only that little sore on the surface. Hectic symptoms come on, and the subject, after suffering untold agony, dies; all from that little sore, that came on the face, or nose, etc., five, ten, or twenty years ago, as a little dry scale. Be ye warned of the lion, and kill him while he is a whelp! I have given the remedies heretofore recommended by various authors, in the treatment of cancer, and need not repeat them here. I occasionally removed cancers, many years ago, with concentrated nitric acid, applying it thus:—I made a cut around the sore with a stiff paste, filled the inner surface with fine saw dust, and saturated this with pure nitre acid (this is a painful remedy), covering the whole with a bandage, or strip of adhesive plaster, let it remain eight or ten hours, then removed it, and dressed with a mild poultice, every six or eight hours for several days, till the slough fell out, then dressed with simple cerate, until well. In this way I have cured several small cancers, that did not return. Anodynes must be given to relieve pain. But when the cancer is large and deep, this remedy will fail. I have treated them in like manner, with the chloride of zinc, and per-chloride of iron, but with less success. Many years ago I used arsenic, both internally and externally, but without any beneficial effects; hence I have long since laid it aside.

For the last twelve years I have used the following remedy in the treatment of epithelial cancer, and where I could apply it, have had no reason to change it. Solid extract of podophyllum one part; chloride of zinc, three parts; starch, half a part; red saunders, or fine saw dust, half a part; water sufficient to form a thick paste. Put the system in good condition, and apply this paste; spread a quarter of an inch thick on strips of cotton cloth, and lay them closely all over the sore, embracing a quarter of an inch margin. Bind this on with adhesive straps; let it remain twenty-four hours: remove it; wash the surface and make a fresh application. Repeat this three or four times, or until the surface of the cancer is hard and white. Then poultice with light bread and water, or slippery elm bark ground fine; renewing the poultices every six or eight hours. The edges of the crust will crack all around, and in six or eight days the slough will fall out. If the margins and bottom of the sore are smooth, you have removed it all, but if there are any rough places left, it is not all out. Then apply the paste on these spots, once or twice, and poultice, and you will remove the whole. Then dress with an ointment, made of equal parts of white wax, mutton tallow, and lard, gently melted together, and stir till cool. It should be washed with castile soap and tepid water every time it is dressed—three times in twenty-four hours, until well. Granulations will spring up rapidly, the wound fill up, and you have a smooth surface, with no puckering, if it is dressed right. During the operation the patient may require some anodyne, as well as aperients, and such constitutional treatment as the general health may require. It requires from six to eight days for a slough to drop out, according to size; and from three to six weeks to fill up and heal over.

2d. Fungoid cancer has its seat in the parent gland cells, which are usually more or less fusiform in shape, and which line the red and white blood apparatus. The cancer cells having their seat in this adventitia, are deeper seated than the epithelial cells, consequently when they aggregate and multiply, they press to the surface, and burst out in a fungus. This fungus being filled with new blood vessels, bleeds at every light touch or abrasion of the surface. The granulations are exuberant, and sometimes of a very considerable size. No matter on what part of the body we find them, they are always of the same character, and have their origin in the same parent cells. The treatment is the same in principle, as with the epithelial cancer, though the remedies applied may differ according to the size of the fungus, and the location of the sore. We shall give them when we come to give cases and their treatment.

3d. The parent involuntary gland cells, which organize the involuntary muscles, is the seat of the fibroid cancer. Hence these cancers do not throw out a fungus at first, but present rather a hard dry sore. Sometimes the edges become everted. They are most apt to be seated on the side over the ribs, or near some solid bony surface. They are often slow in their progress. They may occur in the mamma, and when located there, they may remain almost dormant for years, before they produce an eruption on the surface.

4th. The hematoid cancer has its seat in the parent connective tissue gland cells. These are situated inside the basement membrane, and outside the parent cell layer that lines the white and red blood-apparatus. These are the cells that originate all the products and tissues; the distinguishing prominent principles of which are either gelatine, chondrin or ostein. From these cells emanate connective tissue, cartilage and bone. Here the hematoid cancer has its seat; hence its rapid progress and speedy termination in death, if not timely and properly treated. This is the old hematodes fungoides.

PREPARATION AND COMPOSITION OF REMEDIES, AND RECORD OF CASES.

Before we give any cases and the mode of treatment, we will give the remedies generally used in each classification of cancer. The object of their application is the same, viz., to remove the cancer cells and diseased tissues. But the remedy in the same form cannot be applied in every case. Hence the necessity of varying or modifying it to suit the case in hand. The question may be asked, why so combine these articles of the materia medica before using them. The answer is, by combining them, we increase their power. You might as well ask why combine charcoal, sulphur, and nitre, to make gunpowder; each one has its own appropriate properties, and why not use them separately? Because a proper combination of them makes the object desired for the use for which it is intended. So with nitro-glycerine, the glycerine, sulphuric and nitric acids combined, make a substance that possesses a power a thousand-fold above either of the simple articles. In the same manner we obtain a power by combining several articles for the destruction of cancer-cells, and diseased tissue, that far exceeds the power of either article alone. They are all active medicines in and of themselves, but with all their activity, they, in many cases, have not power to destroy cancer-cells, especially when they are impacted and deep-seated in the fibrous tissue; hence the necessity of so combining the remedies as to give them not only power, but if I may so say, an affinity for those morbid growths that we wish to destroy.

With this explanation I shall proceed to give the remedies and their compounds. That they may be the more easily recollected, I will number them, as far as the general remedies are used, so that when a number is referred to, the exact remedy used may be known, and applied in each case and class.

No. 1. Solid extract of podophyllum (the root), one part; pure chloride zinc, three parts; starch, one-fourth part; red saunders, one-fourth part; water sufficient to form a thick paste. The object of the starch is to give tenacity to the paste, and the red saunders to give porosity, that the plaster may not run, and at the same time give porosity enough to let the full effect of the active articles pass to the sore, where they are rapidly absorbed. This preparation will keep any length of time in a glass or porcelain cup.

No. 2. This is simply a saturated solution of chloride of zinc. It must be chemically pure, kept in a glass-stoppered bottle, and when applied, a glass brush should be used.

No. 3. This is a paste like No. 1, with the exception of using carbolic acid instead of water.

No. 4. This is an arrow made of pure chloride of zinc. Take enough starch to absorb the moisture of the chloride, work together on a pill-plate with a wooden spatula, just enough starch being added to make a stiff paste. Roll the paste into thin cakes, and cut the arrows to a point; then roll gently, and dry on a plate at 212° Fahrenheit. These arrows, properly prepared, and put into a glass-stoppered bottle, will keep any length of time.

These are the potent remedies generally used in open cancer; all others are only auxiliaries, but, in their place, of great importance. We shall give them when we give cases, etc. We shall only give a few cases of each class, deeming that sufficient to guide any one who wishes to try these remedies. We have classified cancer, and shall not repeat the classification.

CASE I. Mr. Webb, August, 1864, called on me with a sore on his cheek, which had existed for several months. Upon examination, I found it filled with cancer cells. This was an epithelioma. The sore was raw, and the edges everted. I made three applications of No. 1 plaster, spread on a piece of cotton cloth large enough to fully cover the·sore and margin; confined it with adhesive straps. It was removed once in twenty-four hours. After the third application the surface of the sore was white and hard. I then ordered a poultice of light bread and water to be applied, and renewed every six hours. In six days the cancer fell out, leaving the edges and bottom smooth. The lump that came out was one inch in diameter, and more than half an inch thick in the centre. By dressing it with simple cerate it healed in due time, and left a smooth scar. It remains well.

CASE II. Major Thomas Shelby came to me about the first of August, 1864, with a sore in his right temple, which had existed for eight or nine years. The sore was as large as a silver half-dollar, dry and hard; somewhat painful at times. I dressed it with No. 1 plaster for four days; then poulticed it for nine days, when it came out, leaving a smooth bottom and edges; dressed it as usual, and it healed kindly, and has not returned.

CASE III. October 30, 1864, Thomas Howard applied to me with epithelial cancer on each side of his neck as large as a dollar. These sores had existed for several years. The edges were everted, and discharging freely a fœtid matter. Made three applications to each, of No. 1 plaster, and poulticed them. In six days they dropped out and healed kindly, and remain well.

CASE IV. Feb. 10, 1865, Mrs. G. W. Lewis applied to me with an epithelial cancer in the left breast, involving the nipple, which was very much retracted. It had been sore for eight years. There was a hard lump in the centre of the breast, and the sore spread all around the nipple, which was retracted. She experienced shooting pains occasionally. General health not much impaired. Made four applications with No. 1 plaster, and poulticed it eight days. It fell out; was two inches in diameter and one and a half inches deep, tapering to a point, in the shape of a cone. The surface on the walls and bottom of the sore were smooth. It healed in due time, and remains smooth and well, and the whole breast soft. Her health remains good.

CASE V. Aug. 18, 1865, Wm. Phelps, aged 67 years, applied to me with an epithelial cancer on the forehead, of nine years' standing.

The sore was nearly two inches in diameter, and reached the bone. The edges were everted, and the discharge cancerous. Made four applications of No. 1 plaster. Then poulticed it, and the cancer dropped out; periosteum removed, bone scaled and exfoliated, but in due time it healed kindly, and remains well. I saw him not long since.

CASE VI. Col. Thomas Bringhurst, of the United States army, stationed in Lexington, Kentucky, July 18, 1865, brought his wife to me with a cancer in her breast. Her age, 29 years. This tumor had been of about three months' standing. It was situated on the outer portion of the left breast. In order to facilitate the action of the medicine, I ordered a small blister over the tumor at night, to remove the cuticle, and cut a little on the true skin. Four applications of No. 1 plaster killed the tumor. I then poulticed for seven or eight days. It fell out, leaving a cavity two inches long, and one and a half inches deep. The walls were smooth and healthy. She then went home to Logansport, Indiana. It healed in due time. Some small lumps afterwards appeared on one edge of the scar, which were easily dissipated by the application of iodine, ∂j; bromine ʒss; lard ʒj. This was applied twice a day for a short time, and all hardness disappeared and has not returned.

CASE VII. 1867. James Clary, aged 45 years, applied to me on September 10th with a fibroid cancer on his neck. It commenced six years ago, as a little blister, and is now four inches in diameter, and deep-seated in the muscles. Edges very much thickened, and everted, but the sore disposed to be rather dry; at least did not matter freely. It was painful. I made four applications of No. 3 plaster. The cancer appeared to be killed; poulticed it, and in eight days it fell out. It was an inch larger than the original sore, and looked well. He left for home in an adjoining county. It healed well, but in six months a small sore made its appearance in the edge of the hair. One application removed this, and he remains well.

CASE VIII. 1867. Margaret Miller, aged 70 years, applied to me September 10th, with an epithelial cancer on the upper lip, embracing the whole lip and the septum of the nose. It was two inches long, had been four years since it made its first appearance, like a little wart.

Four dressings with No. 1 plaster killed it, and it dropped out in six days. The sore healed kindly, but in a few months there appeared a small sore in the corner of the mouth. A few touches with No. 2 removed it, and there has been no appearance of it since, that I have heard of. She lives distant 140 miles.

CASE IX. 1866. Mrs. Colonel Estill, 45 years old. Cancer of the right breast. It commenced four years ago as a small lump in the mamma between the nipple and the axilla. It enlarged slowly, for three years, until it involved two-thirds of the mamma; then ulcerated on the top, when Dr. JAMES BUSH, of Lexington, Kentucky, extirpated the whole breast. In three or four months it healed, but reopened again in a short time; the ulceration extended rapidly, and soon the axillary glands became involved. In this condition she came to the city, and I was called to see her. Her general health was tolerably good. She was cheerful, but reconciled to her fate, believing she could not recover. The sore on the breast was six or seven inches long, five or six inches wide, the pectoralis major gone, and the pectoralis minor fully exposed, and covered with a gelatinous fluid, which emitted a decided cancerous odor. The edges of the sore were greatly everted, and the axillary glands swollen as large as an orange, of a purplish color, and completely filled the whole axilla. This tumor was very tense. There was also a tumor on the right side of the abdomen as large as a hen's egg. The right arm was swollen to twice its natural size. The elbow and shoulder-joints were completely anchylosed. She menstruated irregularly. Her appetite good; digestion good. She slept well. Her mind cheerful. She had lain so long in a helpless condition, that bed sores had formed on the back, but these she thought of little consequence, and I did not examine them. I felt that the probability of a cure was almost hopeless. Yet as she was heart-whole, I determined to do all I could to mitigate her case; I therefore prescribed :—

    ℞.—Acid. acet., - - - - f℥j.
    Aquæ font., - - - - Oj.   M.

Keep cotton cloths constantly wet, and applied all over the sore.

Also to take iodide of potassium, five grains three times a day, dissolved in comp. tinct. uva ursi. It is Huxham's tincture, with the addition of the uva ursi. Signa; dose a teaspoonful three times a day, in a little water. In a few days the cancerous smell disappeared. I then injected the raised edges around the sore with acetic acid and water, equal parts: also the tumor under the arm. This I repeated three or four times. The edges began to shrink, and the tumor under the arm began to shrivel, and to lessen. I continued the wet cloths, and injected the tumor on the abdomen. This soon opened, and discharged a great many cancer cells. The edges of the sore rapidly shrunk away, and it began to granulate, and look healthy. The arm was still swollen, and the joints anchylosed. I now began to feel that it was rather a dangerous experiment to remove the tumor in the arm-pit. The axillary artery might give way, and the patient bleed to death. I used the precaution to have the strongest styptics at hand, if the artery should bleed. I now used the perchloride of iron, by keeping a cloth wet with it, and constantly applied over the tumor. By this means I succeeded in removing the tumor by thin sloughs, until some large blood-vessels gave way. The bleeding, however, was soon checked, by the styptics, and no hemorrhage ever occurred afterwards. I continued the acetic acid to all the open sores. I now gave her in addition to the tincture above, cicuta in the form of pills, two grains three times a day. In the course of three or four months, the whole tumor under the arm was entirely removed, and the sore healed. The great sore was healing kindly. I now dressed it with simple cerate, rendered hard by the addition of mutton tallow; but still washed the sore with the acid. She now took chlorate of potash in large doses, internally, and used a more generous diet. After some weeks she took the iodide of iron internally. About this time the arm became dropsical, and swelled to a great size. By frequent punctures with a needle, the water escaped drop by drop, until the swelling all subsided. The sore filled up, and healed rapidly, but the anchylosis still continued. Now erysipelas set up in the arm; the embolism of the blood-vessels caused this inflammation. I however soon subdued the erysipelas, and the swelling all left the arm. The embolism was removed. The sore gave evidence of soon healing. Cicatrization was going on finely. But now appeared another difficulty. A pain passing from the lumbar region down the left side and thigh in the course of the great sciatic gave her much annoyance. After trying several remedies for neuralgia without relief, her mother suggested that a bed sore might have something to do with the pain. I examined the sore and found a dry hard scale on the left side of the lumbar vertebra, as large as a dollar. It felt hard, and did not suppurate. I made some applications and ordered a poultice. In the course of eight or ten days it cracked around, loosened and maturated. It finally sloughed out, and I found that its point had reached the vertebra at the joint and seat of the spinal nerves; the pain ere this had begun to pass up the spine. I now felt afraid that the spinal marrow was involved, by inflammation, and predicted that it would prove fatal. The pain continued to extend up the back; pains in the head came on, and coma ensued, and she died. The great sore had nearly healed. I believe this to have caused her death, and that she did not die of cancer. I give all the circumstances, and set this down as a fatal case.

CASE X. 1865, September 27th. D. M. Parrish called on me to treat a cancer on the cheek, situated near the nose, on the right side. It was of fifteen years' standing, and had been partially healed several times. It was now dry on the surface, and evidently of the fibroid class. The edges were everted. He is a botanic physician, and understood well the nature of his case. I dressed it with No. 3, and made four applications, when the tumor fell out. There were some fibres left in the bottom of the sore. I made one more application, and then the slough came out in a few days. There were yet a few fibres left, sticking to the periosteum. I touched them with No. 2, and they came out in a few days, bringing the periosteum with them. The inner surface of the sore was now smooth and healthy. It healed kindly, and remained well.

CASE XI. 1865. Mrs. Eliza S. Harland, of Frankfort, Kentucky, applied to me August 25, with an epithelial cancer in the palm of the right hand. She was 59 years of age. Her mother died of cancer.

I dressed it with No. 1 paste. Three applications were made, and it came out in due time, as large as a musket ball, perfectly round and smooth, leaving a smooth cavity. It healed kindly, and remains well.

CASE XII. 1865. Colonel Randolph Bailey came to me June 2d, with epithelial cancer on the cheek. This was an oblong sore, one inch long by three-quarters wide, with some extended branches down the cheek. Dressed it with No. 1. plaster four times, and in seven days it came out, healed, and remains well.

CASE XIII. 1865. May 29. J. H. Van Pelt came to me with epithelial cancer on the nose, of nine years' standing. I treated it in the usual way, when it was cured.

CASE XIV. 1865. Miss Mary A. Magowan came to me with cancer on the right cheek. I dressed it with No. 1 plaster. It was cured in due time.

CASE XV. 1868. This lady came to me February 26th, with a cancer on the left thigh, immediately under the great trocanter. Four dressings, when it came out, healed kindly, and never returned.

CASE XVI. 1864. Edward Oldham, Esq. Small cancer on the cheek, about three-quarters of an inch below the margin of the eyelid. It was fibroid. I dressed it with No. 3, four times; and when the slough came out, as large as half a dollar, it extended under the eyeball, and removed at least one-half the depressor oculi muscle, so that the globe of the eye was partially exposed, but leaving the conjunctiva sound. A small strip of the eyelid remained with the ciliary band and eyelashes. I kept this up by adhesive strips, till the granulations filled the cavity, and saved the appearance of the eye. It healed well, and left very little scar, and a perfect eyelid. Remains well.

CASE XVII. Mrs. Allen, aged 58 years. Had borne four children. Had not menstruated for nine years. One year ago she felt pain and soreness in the small of the back, and outside of the thigh. Shortly afterwards she felt soreness in the mouth of the womb, which rapidly increased. It was examined by her physician. The os uteri was found to be ulcerated. It was cauterized. Since which time she has been under the care of several physicians; been cauterized frequently. I examined her first in November, 1866, and found the entire os ulcerated and everted. The sore was hard, with a rough uneven surface, as if it had been nibbled out by some small animal. Her general health was rather impaired. She had frequent hemorrhage, and very offensive discharge from the vagina.

Here the mode of application of the remedy had to be changed. I have a piston made of hard wood, with a small nob on the small end, and a head to fit the flared outer end of the speculum—a glass speculum, varnished on the outside. Upon the small end of the piston, I bind a bunch of cotton wool. Open the end and place in it ʒss. of the chloride of zinc; close the cotton over it tightly, and saturate the whole with dilute carbolic acid. Pass the speculum into the vagina, and embrace as much of the sore as possible in its end. Pass the piston in, and press the cotton tightly against the sore. Confine the whole with the **T** bandage. Let it remain twenty-four hours. When removed, the surface of the sore will look white, and feel hard. Wash the vagina out with tepid milk and water, and apply the medicine as before. Repeat this four or five times, then wash the surface by injections with tepid milk and water, three or four times in twenty-four hours. In eight or ten days the slough will come away. This process left a smooth healthy sore, which healed in three or four weeks, and to all appearance she was well. But time proved the contrary; after a time a small ulcer appeared on one edge of the cicatrix. Two applications removed this. She went home with the promise to return if anything untoward should appear. She did not return, however, but I visited her once, and found a small weeping sore in one corner of the os, which I thought came from the cavity of the womb. I never saw her again. She failed to visit me, and my business would not admit of my attending her at so great a distance. Some time after she died. I give this case in full, because it proved fatal. I have cured more than a score of similar cases, by exactly the same mode of treatment, with the addition of, in some cases, injecting the womb with a weak solution of iodine for that weeping discharge, after the sore was removed from and healed on the cervix. In every case, however such constitutional remedies were used as the case indicated. In all cases a strict diet must be enjoined. During my residence in Louisville, from 1834 to 1856, I treated as high as eight or ten patients at a time with lupus, epithelioma and fibroid cancer, of the os and cervix uteri, with the loss of not one in twenty. I invented, and had made, instruments for the application of medicines to these parts.

CASE XVIII. 1865. Mrs. Catherine Gay, aged 69 years. Cancer eight years in the lid of the left eye, below, and extending to the inner canthus of the eye, involving the lachrymal duct. The edges everted and showing cancer-cells. This was a fibroid. I dressed it first, March 21st, and covered the ball of the eye with ising-glass plaster, and applied No. 1 plaster. I dressed it three times, and then poulticed it. It came out in due time and the sore healed kindly. In five or six months there appeared a small protuberance between the eye-ball and the nose, not larger than a small pea. I covered the eye ball as before, and applied No. 2 with a glass brush, two or three drops at each time for three days, and then poulticed it. It came out the size of a quill, and three-quarters of an inch long, tapering to a fine point. It healed kindly. The eye perfectly sound, and vision good. It remained well.

CASE XIX. 1866. Roy Stewart, aged 73 years, with cancer on the left cheek, of fourteen years' standing. It was cut out twelve years since, and healed up, but reappeared in two or three years, and progressed slowly, and now shows an abundance of cancer cells of the fibroid class. It was dressed January 20th, with No. 3, four dressings; and in due time it came out. It healed kindly and remains well.

CASE XX. 1865. John Patterson, from Illinois, came to me with epithelial cancer on the right cheek, of five months' standing. I dressed it with No. 1 plaster four times, and in due time it fell out, healed kindly, and has not returned. Heard from him a few weeks since.

CASE XXI. 1866. William Conn, aged 82 years, with epithelial cancer on the hand, between the thumb and forefinger, of five or six months' standing. The sore was dry, with edges everted. Dressed it four times, when it came out. The cure was perfect.

CASE XXII. 1866. Mrs. Julia Forshea, aged 63 years, came to me with fibroid cancer on the left cheek, near the nose, of three years' standing. It was nearly as large as a half-dollar. The edges were everted, sore, dry, and hard. Dressed it four times with No. 3 dressing. It came out clear, and healed kindly.

CASE XXIII. 1866. John Cartney, aged 60 years, came with cancer in the mouth. Fourteen years ago there was a lump on the right gum. His dentist told him he did not like the appearance of it. His family physicians treated it from time to time, but failed to cure it. One of them accompanied him to Philadelphia and New York; but he could find no surgeon that would undertake the cure. He returned home, and made a visit to the celebrated Vermont Springs, but received no benefit from them. On his return home he sent for me. On examination I found the whole of the gums on the right side, below, cancerous, which extended from the symphisis of the chin to the condyle of the lower jaw and the parotid gland and submaxillary glands; and all the lymphatics on that side of the neck were involved and swollen; there was a fissure more than an inch long separating the gums from the alveolar processes, and quite half an inch deep. His general health was broken down; slept but little, and when he did sleep his mind was wandering; appetite bad; digestion very much deranged. I told him I could not cure him, and refused three separate times to treat the case; but the entreaties of himself and friends overcame my better judgment, and I promised to do the best I could for him. I told him there was only the shadow of a hope, and that was the smallest thing he could think of. But I must try. This was a fibroid cancer. I could not use the ordinary remedies in this case. I gave him constitutional remedies, and such local applications as were admissible in such a case. I reduced the enlargement of the parotid and submaxillary glands; but his teeth all came out, and many large strings deeply seated under the tongue, but the sore extended in the cheek and toward the chin. He continued, however, to have great pains in the head; his bowels became very much deranged from constantly swallowing the saliva and secretions from the sore in the mouth. He lingered for eight or ten weeks, and died. Had this case been properly treated in time, he might have been cured, as the following case will show.

CASE XXIV. 1861. Henry Lancaster, aged 51 years. He stated that about February of that year there appeared two small specks on the left side of the tongue, of a rather yellowish cast. "They burned like fire," to use his own language. DR. S. L. ADAMS was then his family physician, and saw them the first day they appeared, but did not regard them as anything serious. They continued to spread and deepen for thirty days. DR. ADAMS said he could not cure them. He then sent for me. At that time the sores had spread and came together, and were about one inch long and half an inch wide, and nearly half an inch deep, with rather a bluish appearance, the edges beginning to evert. The cells were to be seen. I pronounced it epithelial cancer. In this case the ordinary remedies could not be applied; but some powerful remedies must be used, or the patient must soon die. I therefore applied the nitrate of copper in strong solution, by making a thin mop on a small paddle of wood and wetting it with the solution, and filling the sore with it. I let it remain a minute or two, then covered it with sweet oil with a similar mop. In order to more easily come at the sore, I grasped the point of the tongue with my finger and thumb, placing a piece of cotton cloth on the point of the tongue to prevent my finger from slipping. In this way I drew the tongue out, and held it firmly till the application was made. I repeated this every day for one week, when, to all appearance, the cancer was killed. I dressed it then with the oil only. In a few weeks it healed; but in a month more there appeared a small lump on the posterior part of the eschar, as large as a small bean. I reapplied the medicine a few times, and this came out in a string at least one inch long, and as large as a quill. After this the sore healed kindly, and remains well. I saw Mr. Lancaster a few days since, and looked at his tongue. It is perfect, and no appearance of disease about it. He says it is perfectly well, and has been ever since I cured it first.

Here was a case of cancer of the tongue taken in its early stage, and cured.

Case XXV. 1868. William Dixon, aged 38 years, with hematoid cancer in front of the left ear. Had it cut out six months since. This tumor was the size of a large orange. It was filled with large blood-vessels; had bled profusely and frequently. After being cut it did not heal, but grew rapidly and continued to bleed freely. I did not encourage but rather discouraged him from having it treated. He, however, requested it, and I applied No. 3 dressing on it three or four times; then used No. 4, by inserting some six or eight arrows all around the tumor, sinking them an inch and a half deep by first opening the way with a bistoury to that depth. I then poulticed it eight or ten days, when it fell out the size of a large orange. It healed partially, but a small tumor sprang up in one edge of the sore, which I removed with No. 3 dressing. It partly healed, and he left for home. I heard he died of debility some six months after.

Case XXVI. 1868. Samuel Parvin, Esq., aged 65 years, with epithelial cancer on left side of the nose and cheek, of five or six years' standing. General health tolerably good. I dressed it with No. 1 plaster, three applications. It came out in due time, healed kindly, and remains well.

Case XXVII. 1869. Mrs. Ann Williams, epithelial cancer on the chin, of eighteen months' duration. She had been treated by a good physician, but not cured. A few applications of No. 1 plaster, and it was cured.

Case XXVIII. 1869. Mrs. E. Lobbin, Frankfort, Kentucky. This lady applied to me with a cancer on the left side of her nose and face. It had been twenty-four years since its first appearance, like a small wart, near the nose, about midway from the eye. This wart, after some years, became a little sore on the top; she had it cauterized, but it would not heal. It progressed slowly for many years; ultimately it became a permanent sore, and burrowed deeper. It was dry, with the edges a little raised. This state of things continued with a slow increase, until the sore was as large as a quarter of a dollar, and had eaten through, into the cavity of the nose, and destroyed a portion of the nasal bone, pervaded the maxillary bone, and broke through to the surface under the left eye. It was as large as half a dollar, and in this condition she came to me. These sores were full of cancer cells. This was a fibroid cancer. I hesitated to treat it, but she insisted that I should do so. I applied No. 3 dressings four times to the whole surface. Poulticed them, and in ten days the sloughs came away, but the bones showed the cancerous influence. By using proper remedies to exfoliate the bones, several large scales came out of the face; from the superior maxillary, the whole of the vomer to the nasal suture, came away. Several pieces of bone as large as twenty-five and fifty cent pieces came out of the face. Then the whole sore healed, and left a cicatrix not so unsightly as I expected. An artificial nose will restore her appearance. Constitutional remedies were used all the time. It required two months to perform this cure, and it remains well, and her health is good.

CASE XXIX. 1866. Mrs. Sarah A. E. Hughes, aged 62 years, fungoid cancer in the left breast, below the nipple, as large as the top of a small teacup, with greatly everted edges; very fœtid discharge of creamy-looking pus. It commenced two years since, in a tumor on the lower part of the breast. Her breasts are very large and fleshy. After one year's existence, DR. JAMES BUSH cut it out. It healed, but soon broke out again. In three or four months it was again a running sore. It has now been open six months. The whole breast is very large and inflamed. After using the necessary remedies to reduce the inflammation, I dressed it with No. 3 dressing. It required a number of applications to destroy the fungus and kill the cells. I then poulticed it. In time the whole lump dropped out, weighing nearly or quite half a pound. The inner surface of the cavity looked smooth and well. She went home, nine miles in the country, and neglected to keep the outer surface of the sore open until it could heal from the bottom. The outer edges of the sore came in contact, and healed before the sore healed from the bottom. You must recollect, this sore was more than two inches deep and about four inches long. This was bad management, and contrary to my directions. Finally, in February, 1868, there appeared another fungus on the outer edge of the old scar, which was removed by the same remedies in due time, and never reopened; but the breast felt soft in every part, and no pain or soreness was felt anywhere. Her health appeared to be perfect. About twelve months afterwards, there appeared a hard cake in the ribs, about two inches below the breast. There was no running sore, but the skin was dry and hard. The spot had shrunken in as large as half a dollar. The breast was soft, not a lump in it, and the axillary glands unaffected. This hardness was very unlike that which appeared in the breast before. The breast is, to all appearance, as well as the other breast. This place on the side did not discharge any fluid of any kind, but gave some sensation of soreness when pressed upon. This was evidently a fibroid. I made an application of No. 3 dressing, and repeated it three or four times, and poulticed it. The lump fell out. There was a small lump left, which was attached to the fourth rib. I was sick, and could not attend her, as she was nine miles in the country; and by persuasion Mr. Hughes employed a man who professed to cure all cancers without fail. He attended to the case one month, and said he could not cure it, but his father could cure it. He was sent for, and attended to the case five or six months, remaining at the house all the time. Instead of treating the cancer, he insisted on taking off the whole breast. Mr. and Mrs. Hughes both declared the breast was sound, and entreated him to treat the cancer; but he declared he could not cure the cancer unless he removed the breast, which he finally did by escharotics and the knife, leaving the sore all this time untouched. By this time the cancer had spread until it was open, four or five inches long, and looked like a deep gash that might have been made with a broad edge-tool. It had extended completely under the pectoralis major and up into the arm-pit, and the axillary glands were all enlarged and filled the axilla; looked blue, and was very tender to the touch, yet the arm was not swollen. The sore now presents a surface five inches long and nearly four inches deep, passing over the second, third, and fourth ribs, and discharging a matter of the true cancerous odor. Her skin is covered with dry, branny scales, and looks as if flour had been sprinkled on it. Her appetite poor, and her digestion very much impaired. Complains frequently of pain in the region of the gall ducts. In this condition HOWARD left her, after Mr. Hughes had paid him $3500, for which he had warranted a cure. Mr. Hughes now came for me again; and in the above-described condition I found her. My hopes now of her recovery were blasted, but it was my duty to do all I could for her. By repeated applications of Nos. 1, 2, and 3, and the use of acetic acid and carbolic acid, I succeeded in removing the tumor from under the arm, and all the surface of the large sore, and it had healed to two inches long and one inch deep, and all healed in the arm-pit. Now the mischief lay internally. A pain commenced up in the region of the ductus communis choledochus, and the secretions from the liver were obstructed for eight or ten days. I now felt satisfied the case would prove fatal. The liver was cancerously affected, which proved my suspicions well founded. She died early in July.

CASE XXX. The following case was presented to me March 8th, 1871. Mrs. T., from Logansport, Indiana. This lady called on me for treatment of a tumor in the right breast,—the history of which she gave me as follows: Some two years or more ago, she felt a small tumor in the right breast about an inch above and to the right of the nipple. The lump gradually enlarged, and felt a little sore on pressure. Occasionally she felt darting pains shoot through it, but this gave her no uneasiness. The lump continued to grow, and the darting pains to increase, especially when the lump was pressed upon. The skin adhered to the tumor, and the nipple becoming involved began to contract. She now, for the first time, showed it to her family physician, and asked his advice. He examined it, and pronounced it malignant, and advised extirpation immediately. Not being willing to have the knife used, she determined to have my advice, (there being a lady of her acquaintance in the neighborhood, from whose breast I had removed a similar tumor some eight or nine years since, and it remained sound and well,) Mrs. T. and her husband came to me, March 8th, 1871. Upon examination I found the tumor in the right breast as large as a duck's egg, hard and adherent to the skin, the nipple involved and contracted, but the tumor loose from the pectoral muscle. She suffered lancinating pains. The axillary glands and lymphatics seemed to be free from the disease. Her general health tolerably good. Age 46 years; has had two children, youngest 14 years old. My diagnosis was a scirrhus tumor of a malignant character. She desired me to remove it without the knife, as I had that from Mrs. Binghers. In order that the medicine might take hold more effectually, I applied a small fly blister over the tumor and raised the cuticle. I then made five applications of No. 1 plaster over the tumor. This produced a crust three-fourths of an inch deep; but I found there was still a portion below the crust that was not killed. In order to save time, which it would require to form another crust, after removing the first one, I determined to inject the deep portion of the tumor, and so facilitate its removal. I passed a curved bistory through the crust, and into the remaining portion of the tumor. This was not painful. I then injected with a slim-pointed gutta percha syringe, about a drachm of fluid—of ten grains of the chloride of zinc to a drachm of water. I was apprehensive of pain, but to my agreeable surprise, the pain was slight, and did not last long. She had an attack of colic that night, which was relieved by a few drops of laudanum. Whether the colic pains were induced by a portion of the acid entering the circulation, I am not able to say; suffice it to say that no untoward symptoms followed the use of the injection, but with the best effects, as we shall presently see. I now commenced the ordinary process of poulticing, and in seven days the whole tumor came out, the crust about three-fourths of an inch thick, and two and-a-half inches in diameter. Attached to the under portion of this crust was a soft lump, more than an inch in diameter, and three inches long, of a pale pink color, and the consistence and feeling of the finest velvet sponge. The surface it left was smooth and even, and not one drop of blood escaped. When the tumor came out, it appeared as if at least one-half or more of the breast was removed. The nipple with all the milk vessels centering in it came away. Constitutional remedies were used during the treatment, and advised to be continued for several months. She returned home, some 300 miles. I have heard nothing from the case since.

CASE XXXI. The following case presents some interest; I therefore give it. Mrs. M., of Covington, Kentucky, aged 35 years, applied to me with an epithelial cancer in the left temple, extending to the eye-brows. It was an inch and-a-half wide, and two inches long. There was a large hard lump in the centre of the sore. It commenced several years ago, like a red wart, and remained stationary for a year or two; then began to spread and "eat the flesh," as she described it. It was removed by the knife, and partially healed, but soon began to corrode the flesh again. It was cut out the second time, but this time it refused to heal. Her physician told her it must be cut again, but she declined to have it done. She applied to me March 27th, 1871. After getting the above history of the case, and examining it to my satisfaction, my diagnosis was epithelioma, the mildest form of cancer. I applied four dressings to it, one each day ; then poulticed it, and in due time the slough came away, but to my disappointment there remained a rough granular lump in the bottom of the sore, as large as an inch of the ring finger. This must be removed. Three more dressings with No. 2, and a thin slough came away. But the tiger still held his grip upon the periosteum, and I then made four applications of the plaster, and poulticed it till it all fell out, leaving only the periosteum to cover the bone. The surface was smooth, even, and not a malignant fibre left. She returned home. I heard from her May 21st, and it has all healed up clear and smooth. I have been thus particular in giving the treatment of this case, because of the number of applications necessary to remove all the cancerous growths. I refer this to the two cuttings it had undergone. The knife, as usual, had not removed all the cancerized lymphatics, hence the return of the disease, and the difficulty of removing every vestige. The best surgeon in the world cannot tell the direction the diseased lymphatics take, hence the return of the disease after the most skilful operation. It requires something that has an affinity for the cancerous poison, and will follow it to its extremity.

I have now given nearly all the fatal cases that have occurred in my hands for the last six or seven years, and a specimen of the cases cured, so as to give a sufficient knowledge of the remedies used. It must be recollected that in almost every case alteratives and constitutional remedies were more or less used. These I have not generally mentioned, as every physician who treats cancer—and no other person should attempt it—well knows what constitutional remedies the case demands. Aperient and alterative medicines are required, according to the constitution, age, and temperament of the patient. I have only given cases enough to show the different remedies I use in the different classes of cancer, though I might fill a small volume with cases. I shall now give a short summary of the work, and close the subject; hoping that some one desirous of benefiting the race and promoting the science of the healing art will take up the subject and improve upon it, and carry out the great object in view in this work, the successful treatment of this insidious and formidable disease,—until a remedy be found that will check its progress, control its action, and save the lives of its subjects. Our Materia Medica is yet incomplete. The Botany of our country is only partially known, and our mineral resources not fully developed. Who can tell what the sciences of Chemistry and Pharmacy may yet bring to light? My belief is, that the God that made us knew all the calamities to which we would be subjected, and He has not left us without a remedy for them all, both of soul and body. And it is our duty as well as privilege to seek after and apply those remedies. To this end we are endowed with mental and intellectual powers sufficient for the investigation and discovery of these means and applying them according to the rules of physiology and pathology. The time will come when cancer will be treated as successfully, if taken in time, as any of the exanthematous diseases.

Delays are dangerous in all diseases; but, taken at the right time and properly treated, almost any disease may be cured. So with cancer. Take it at the right time, treat it according to its proper classification, and no more mortality will attend this than any other disease of a grave character. There is one thing yet to be ascertained, and that is—What is the true nature and chemical character of the fluid contained in the cancer cell? He who will demonstrate this will have crowned his fame with a wreath of honor equal to that of JENNER when he discovered that variolous matter passed through the cow would so modify small-pox as to render it perfectly safe and innocent, and prevent the subject from taking the original disease.

He who truly ascertains the chemical character of the cancer cell, will open the way for the antidote to that poison, and internal as well as external cancer, will become one of the manageable diseases of the human race. What LIEBIG of the world will achieve that immortal honor? I do not presume to say, that I have, in this little work, arrived at or near the zenith of investigation, but I do say and claim, as an undoubted fact that I have classified cancer, in a way I have not seen done by any other author, and this classification is founded upon strictly physiological and pathological principles. I do not say it cannot be improved, but I feel that I have laid a foundation upon which some other man may build, and if he should take away some of the structure, and place a better support, and add some new basis to the building, and introduce some new materials for its completion, I shall not object, so that sufficient reasons are given for this addition or "change of base." I lay it down as a fundamental fact, that cancer is always to be known by the presence of its specific cells, and without these cells, there is no cancer; and wherever they are found, the case is cancer, class it as you may, or class it not at all. I also state that cancer-cells are specific, persistent, and percolating, that they are always local in their first appearance; and that they never become constitutional except by absorption,—that cancer is not a blood-disease, as it was once believed to be by the profession, and he who yet believes it gives evidence that he is far behind the progress of the profession in physiology and pathology, and cannot be relied upon to treat this disease. I have also shown clearly, I think, that tubercle and cancer are not the same disease. That tubercle shows no cancer-cells—it is only born to die, while cancer-cells are born to propagate. I have shown also that scrofula is a distinct disease from cancer, and that the treatment for the one will not cure the other. Moreover, scrofula is a disease of youth, while cancer is a disease of age.

I have numbered the principal remedies for external application, used in the treatment of local cancer, for the convenience of the practitioner. These are the principal remedies used, not however to the exclusion of all others, for in many cases, especially in those of long standing, you must use constitutional remedies. These, as a general rule, must be left to the judgment of the physician, for whom alone this work is written. I have purposely given the remedies used by the most scientific physicians and surgeons of the medical world, so far as they have come to my knowledge, that those who have not read them may have a choice of remedies. I have given mine, after trying those I thought best. They are the result of fifty-five years in the profession, adopting them after trying a great variety of others, and they have been used in hundreds of cases. The reasons for my compounds may be questioned, but reflect for a moment, and you will see that by compounding substances not potent in themselves individually, they become the most powerful agents known in chemistry and to pharmaceutists. I claim that these compounds have more or less affinity for the morbid cancerous structure, more than they have for sound structures. Apply your mind to these subjects, and you may improve on my labors as I have on the labors of others, for I have not neglected to avail myself of the researches and experiments of those who have gone before me, or are cotemporary with me. You do the same, with this little book, and take another step forward in the science of our noble healing art.

With these reflections, I leave this work to my co-laborers in the field of Esculapius. I have made it a rule all my life not to dispute a point I do not understand, nor to condemn an opinion that I have not examined, nor ignore a subject that I have not properly investigated. I know with man "time is short, opportunities of knowledge rare, experiments fallacious, and reasoning difficult." I also know that close research, connected thought, and perseverance, will overcome many difficulties. I have one more word to say, or shall I repeat what I have said above? We lack the analytical knowledge of cancer-cells. He who makes this discovery, and shows exactly their constituent parts, will justly be entitled to the immortal honors of a world-wide benefactor. When we know these facts, we shall be able to adapt the remedies to their destruction, and treat cancers developed on internal organs, with that success that will crown our profession with immortal honors. Who will make the effort, and gain the immortal crown?

VIEWS OF VARIOUS AUTHORS ON CANCER.

To show that the medical profession has not been negligent in their attention to this formidable disease, we will take a retrospect of the opinions and practice of some of the most eminent physicians and surgeons of the past and present age, in addition to those already mentioned in this work. These authors are not confined to one country, but are found in Europe and America. The true physician wearies not in well doing, but is ever ready and willing to labor for the preservation and benefit of the race, burning the midnight taper, and wearying the active brain to burnish the sword, and wield the remedy that will lay low the fell monster that slays its thousands of our noble race. We may omit to notice some who are of equal reputation with those mentioned in this brief review, nevertheless when their writings are examined, they will be found in all the fundamental and essential points, to mainly agree. This we do that the profession may have in a concentrated form, the previous as well as present opinions and practice in this formidable and too little understood disease. Nevertheless, the profession [is alive to inquiry, research and experimental effort, that peradventure the true nature of this disease, and the remedies therefor, may be discovered. God speed the day.

The following is the diagnosis of cancer and cancroid growths, by DR. MORRIS H. COLLIS, surgeon to the Meath Hospital, from the *Dublin Hospital Gazette*, Oct. 15, 1855, p. 276. This will show the interest felt at that time by this eminent surgeon, in the diagnosis and treatment of cancer. He says: " Epithelioma commences on the skin in a dry and scaly wart, which falls off from time to time, and exposes on each occasion an increasing wide surface of ulceration. At the junction of skin and mucous membrane, it begins as a chap or fissure with indurated edges, and has a similar tendency to scab and ulcerate. When it originates on a mucous surface, it is either as a flat condylomatous wart, or as a deposit in the mucous follicles. Thus it is always as a local hypertrophy of the investing epithelium that it commences; and from first to last, it frequently retains this strictly local character. Many of these growths upon the skin are nothing more than hard horny masses of dry epithelial cells, which occasionally fall off, and are slowly replaced. These may be either prominent or flat. We have frequent examples of them along the edges of the lip, upon that part which is covered with colored skin, external to the line of moist membrane. These are very harmless, and may last for many years' unaltered in extent, or depth, until accidental violence or inflammatory action excites them to spread. However, in the majority of epithelial growths, ulceration of the integument, and deposit of epithelial cells in its substance, takes place sooner or later. The skin splits into papillæ, as in a wart, and the ever-increasing deposit cells upon the surfaces of these papillæ separates them more deeply and more widely. As these cells are wholly extra-vascular, and dependent on percolation for what fluid they draw from the economy, a process of disintegration is continually going on in them, and the surface of parts not subject to evaporation, is soaked in the putrid remains of cells. How far this aids in the spread of the disease by exciting inflammatory action and ulceration of the sound structures, is hard to say. On mucous membranes this tendency to ulcerate is very marked; cauliflower growths from the uterus, for example, although prominent on the surface, and consisting of hypertrophies of the epithelial cells, have their seat on ulcerations, which may destroy large portions of the organ. The ulcers, when exposed, present jagged, irregular margins, and surfaces; enlarged papillæ are visible during the early period of the disease, when the surface is wiped free from the discharges. At the edges they may be detected at any period. The neighboring glands are affected irregularly; sometimes not at all; sometimes at an early stage.

"Upon the whole, the impression on my mind is, that the tendency to infect neighboring glands is greater when the disease has its seat on the mucous membrane; and least when it is on the skin; but no positive rule can be laid down at present, as to the local infection. General infection must be extremely rare; for those who oppose the classification into cancer and cancroid, can bring very few satisfactory cases of general poisoning from genuine epithelioma. The disease often kills by local destruction, and general waste, but without anything which can be called special or peculiar cachexia. When removed even completely, it will sometimes return in an aggravated from, and with rapidity; but this is quite exceptional. In the majority of cases in which the disease is removed completely, it either does not return, or does so slowly after a very long interval. All these facts show that there is much less tendency to general poisoning than in cancer, and justify us in classing it apart, especially as it is found to possess a different anatomical structure."

But suppose these epithelial sores are found to contain the true cancer-cells, why attempt to draw the distinction between them and true cancer? What is it that points out the true cancer? Is it not the true cancer-cell? Then, whether the sore be epithelial or otherwise, if it possesses the true characteristic of cancer, the well-known cells, be it large or small, deep-seated or only superficial, it is cancer, varying only in type, and classification. The author goes on to say, "Another class of cancroid tumor, to which the general name of fibroplastic is rather incorrectly given, includes several tumors of rare occurrence, and with slightly diversified peculiarities. The fibro-plastic cell is not the special element of any tumor of a cancroid nature, but the elements of these fibro-plastic growths resemble it more than any other cell. The fibro-plastic cell is found in all lymph which is undergoing the process of organization, and is consequently often found in tumors of any kind. The class of growths to which I now allude, are called by Mr. Paget, fibro-nucleated, recurrent fibroid, etc.

"The microscopic elements of which they are composed are small, and very pale. They vary in diameter from the 1.1500 to 1.2200 of an inch. They contain a small nucleus, and have a tendency to become elongated into perfect fibres, arranged in a radiated manner, and little organized. There is no juice obtained from a section of these tumors; and although they bleed freely when cut, it would seem to be rather from some want of power in their arteries to contract, than from the number or size of the vessels contained in them. If squeezed or torn up, they break readily into little acini, which are individually remarkably tough and resistant. The minute fibrocells appear to radiate from the point in each acinus, by which it is attached to the rest. The tumor which is situated under the skin, and I believe always connected with it, is firm, moderately elastic, irregularly lobulated, moveable among the tissues and solitary. It does not affect the neighboring glands, except by direct contact through the lymphatics. Nor does it produce general cachexia, independent of waste as cancers do. Its strongest and unaccountable characteristic is its frequent and rapid occurrence after the most complete ablation, even after amputation at a considerable distance from the original seat. It invariably returns in from three to six months in the cicatrix, and requires repeated removal, which therefore should always be complete, but very sparing of the healthy tissues, and repeated early on the reappearance of the growth."

The author goes on to describe many other sub-varieties of both cancer and cancroid. The main features of both classes of morbid growth, he endeavored to lay before his class, as fully as was consistent with brevity. He then lays before them some rules and remarks on diagnosis of tumors, with regard to a correct guidance in the important points of operation and prognosis. "When a patient is presented to you with a tumor, you must note, first, certain circumstances relating to the individual. Second, such properties of the tumor as manifest themselves to your senses: Third, the effects of the growth upon the patient. Then comparing these with each other, and with the anatomical structure and known character of the various growths, you will eliminate such as are incompatible with your examination, and thus arrive at what the tumor is, by ascertaining what it is not. First, then, you will note in each such circumstances about the individual, as age, sex, history, etc. As to age—infancy and childhood are liable to acute cancer, enchondroma, and strumous swelling. Adult life is the period for all hypertrophies, such as fibrous and fatty tumors; for the cancroids, and for the acute cancers, in organs which are late in arriving at maturity, as the organs of generation. As the patient advances in age, slow growth in epithelioma and scirrhus becomes predominant. Cancer and cancroid, form the vast majority of the tumors that are met with, if age be excluded from our consideration.

"Sex exercises still greater influence on our diagnosis. When we glance at statistical tables, we find that cancers of the female breast, and of the uterus, surpass in numbers all other tumors, in all situations put together. Of these scirrhus of the breast and uterus, form the bulk. We have a small proportion of acute cancer of the uterus, and smaller still of the breast. Women are also subject to innocent mammary tumors of the breast. Both sexes are subject to epithelioma of the rectum and genitals; and men more generally than women, to epithelioma of the lips and face. As regards temperament and appearance, thin sallow, unhealthy-looking people are supposed to be more subject to cancer, but the lardaceous variety shows that the fattest people are not exempt from it. A melancholy brooding disposition is often connected with cancer, as it is with any disease of perverted or deficient nutrition; but on the other hand, whatever amount of cheerfulness of temper may be an item in favor of recovery, after operation, it is little protection against primary development of disease. There are sufficiently numerous and remarkable cases on record of both cancer and cancroid occurring in successive generations to prevent our denying the influence of their hereditary transmission; but at the same time, these are few in comparison to the total number of cases. He says he has seen curious and marked instances of the repetition of encysted tumors, which have just as much weight for hereditary influence, as the facts adduced in its favor, in the case of cancer. Accidental violence is in much the same category; it may give an impetus to a latent tendency, and a start to the deposit, and no more. It seems to me very doubtful whether the statement is correct, that the adenoid or common mammary tumor has its origin in effusions of blood.

"We now come to the properties of the tumors themselves. On this point we must note the situation, size, weight, shape, and mobility or fixedness of the tumor; its duration and rate of progress; the color and alteration of the integuments; the appearance of the ulcer, if one exists, and the nature of the discharge; the solitary or multiple nature of the tumor; the glandular irritation or infection in the neighborhood, or at a distance. They must all be carefully examined into and weighed. Acute cancer is rapid, and attains a very large size; is heavier than any non-cancerous growth; is seldom multiple, implicates in time all neighboring structures, though it seldom poisons the glands when primary; and it almost always returns rapidly and extensively in the glandular and venereal organs after removal"—so says the author quoted: "Adenoid tumors are much smaller, very chronic, very movable, perfectly harmless, and never affect the glands, or the system, though they are sometimes multiple in the same breast, or in both breasts. Scirrhus is small, of stony hardness, drawing in the skin and all the fibrous tissues into itself, and finally the nucleus and every other structure. Cancer in any shape does not ulcerate early; this distinguishes it from epithelioma as the preceding symptoms do from adenoid, fatty, or fibrous tumors, and from the recurrent fibro-nucleated. Fatty tumors are easily known by their lightness, and largely lobulated surface and scolloped edges. The most puzzling of all its physical characters is the cystic tumor of the breast, when combined with cancer as it frequently is; except in the plainest cases it is impossible to distinguish by the hand and eye. The presence of cysts may be readily detected in many instances in which the cancerous deposit is small in amount and veiled by the fluid contents of the cysts. Alterations of the integuments, and the general symptoms, must guide us to the discovery of cancer; and in the other case, if we are sure of the presence of cancer, we need not be thus careful to diagnose the existence of cysts. I have seen simple cysts, the result of varices, mistaken for small fatty growths, and I do not know any certain means of distinguishing them. Fortunately, a mistake here would be of no practical importance, as both tumors are harmless. To overlook cancer, in combination with cysts, would be more serious. The last points to be considered in relation to our diagnosis are the effects of the growth upon the individual; these are pain, cachexia, wasting. Pain we have already spoken of. Cachexia is of value in the diagnosis of tumors, if it appear before they ulcerate, as it generally does in cancer. In other diseases, inclusive of the cancroid, it does not appear until the patient is worn and wasted by discharge, and loss of blood by deglutition, or absorption and decomposition. Wasting may be included in these remarks on cachexia.

" With regard to operation. Encephaloid is more often removed with temporary, or even permanent success, than scirrhus. Success, however, is very rare, and when the cancer returns, it is with extreme violence. In the young you will give every chance by removing early and freely. A respite of years from suffering, if not recovery, is in a few cases secured, and where a premature and horrible death would otherwise be certain. The duty of endeavoring to secure this change is plain. In the old, or even in the adult, you have less hope of success; still, operation is justifiable and proper when the powers of life are not depressed, and the tumor easy of complete removal, without such loss of integument as would delay recovery from the operation. Scirrhus, though often removed, is seldom, if ever, permanently cured. When it is adherent to the deeper parts; when the lymphatic glands are the seat of deposit; when the glands of the skin are the seat of small disseminated beads of cancer, or when its interstices are infiltrated, as in the lardaceous form, it is worse than useless to operate. Relapse is certain and rapid. When the growth is very chronic, and very small, in a weak and atrophied individual, and the chances are that an operation will convert into an active agent of destruction a comparatively harmless tumor, which, if let alone, would have lasted for years with little local alteration, and little injury to the patient's life, in such cases removal does no good, and may do much harm. In epithelioma the general rule is to operate if you can remove all the tumor. Relapses often occur locally, but does not prevent a second or third removal. When the disease originates in a mucous membrane (as in cancroid) of the scrotum, genitals, or rectum, the tongue or gums, you have little chance of permanent cure. In many of these cases, the addition of the actual cautery to as complete a removal as possible, is a valuable aid. The cicatrix of a burn is not a favorable nidus for epithelial growths, and, by its great contractility, it seems to check the hypertrophic deposit in its neighborhood. With regard to recurrent tumors, I have already said they must be removed as often as they appear, and with as little loss of healthy structure as is consistent with their complete eradication." Thus we have the theory, prognosis, and remedies of Dr. MORRIS H. COLLIS, who wrote on this subject in 1855.

The subject of cancer at that day so far engaged the attention of this great man, that he seems to have spared no labor or pains to ascertain its true character, and he clearly discovered that there are two classes of this disease, and prescribes his remedies accordingly. Much credit is due to him, and although he frankly acknowledges his failure to cure the disease, yet he shows an inquiring mind on the subject—and this is sufficient to prompt us to still further research and redouble our efforts to find a remedy that will be still more successful. The physician who sits at his ease, without an effort to advance the science he has chosen for a profession, should close his library and engage in some other business. Let not this be said of our noble profession, but let us take all the light that has been given to us, and by judicious and thoughtful effort reach forward, grasp, and develop other truths, bring them to bear on the subject, and so prove ourselves the true philanthropists of the race. He who studies the profession simply for the pecuniary gains he may make by it, at the expense of human life, using only the effusions of other men's brains, and not using his own, is not worthy the name of physician or surgeon. Let not this accusation lie upon any of the noble sons of the healing art.

On the diagnosis of cancer by Dr. ALEXANDER HENRY, the question as to the utility of the microscope resolves itself into two points:

First, as to the absolute value of the instrument in diagnosing a malignant tumor; second, as to the existence of specific cancer-cells. On the first point we have several conflicting statements: First, that in some tumors of an undoubtedly cancerous nature, cells are absent; second, that they have been found in tumors of a non-malignant character; third, that the absence of cells from a hard tumor shows it to be non-malignant. On these statements I can only briefly remark, first, that in doubtful cancerous tumors, as shown by their softness, it is possible that the cells may have undergone a process of disintegration, though I doubt whether cells could not be found in some portion of such tumors; second, that cystic sarcoma,—to which I suppose reference is made when it is said that cells are found in non-malignant tumors,—has a great tendency to degenerate; third, that we must remember that apparently simple tumors are known sometimes to assume a malignant type, and therefore I scarcely see how it can be predicated of any one of them that it would not act in this way. After all, as I shall presently take occasion to point out more fully, the question of malignity is a relative one, and as cells are most liable to assume that state of action to which I would apply the term malignant, I think that the existence of cells in a tumor affords grounds for regarding it as either malignant or likely to become so, while the negative evidence is only valuable so far as it shows the most obvious conditions of malignity not to have yet been assumed. As to the specific nature of the cancer-cell, I can only state here that a consideration of the opinions of the numerous microscopists to whom I have referred, together with the few observations I have been able to make personally, lead me to doubt whether there is a diagnostic cell. I should be guided more by finding cells in situations where they ought not to be; and if there is any cell more diagnostic of cancer than another, it is the large parent-cells with from three to five smaller ones in it. But the absence of such cells does not show that the disease is not cancer." [*Associated Journal*, Dec. 15, 1855, p. 415.]

Here is still another evidence that the profession is giving attention to those distinctive characteristics of cancer, and although there were differences of opinion at that day in relation to the distinctive character of cancer-cells, and that by men of undoubted ability, yet the question remained to be solved until the immortal BENNETT and WALSH, confirmed by those never-tiring German physiologists and pathologists, have settled the question beyond a successful refutation, that all cancer tumors or sores are to be designated and determined by their specific cells. When these cells are found, whether the disease be epithelioma, fungoid, fibroid, or hematoid, it is cancer, and, if let alone, will sooner or later destroy the life of the patient.

SPROUTING CAULIFLOWER CANCER.

Samuel Solly, Esq., F. R. S., Surgeon to St. Thomas' Hospital, writes as follows on this disease:—"This form of cancer has been called by some writers, the soft cancer, to distinguish it from scirrhus or hard cancer; by others, carcinoma medullare, or brain-like cancer; fungus hematodes, or bleeding fungus. It is that form of malignant disease which is more certain to return after an operation than any other. Notwithstanding this, it is often our duty to operate, not with the hope of saving life, but with the certainty of removing a loathsome mass, which makes life wretched. In many of these cases, the operation prolongs life, though not to the natural period of man.

"The rule which I have found, on the whole, the most safe and judicious in regard to the amputation of the breast for malignant disease, is this: to operate in all cases in which your advice is sought in the early stages of growth, before the surrounding glands are implicated, and the patient's health so much impaired as to render the immediate effects of the operation dangerous; to operate in cases where the growth of the disease is more rapid than the inroads upon the health of the sufferer; where a loathsome mass goes on sprouting, bleeding, ulcerating, discharging, to the infinite disgust and distress of its victim; where there appears no prospect of death putting a speedy end to her agony, and where the powers of life are so unequivocal that the operation does not threaten any immediate danger." He says: "The cases in which I would not operate are those cases of scirrhus (stony cancer) which have been advancing slowly, silently, and painlessly for years, without much observation or anxiety on the part of the patient, and without any knowledge on the part of the friends; where a mass has gradually been formed implicating not only the whole mamma, but also the glands in the axilla, and sometimes the glands above the clavicle, attached to the pectoral muscles, and even to the intercostal muscles and ribs before the surgeon sees it at all. In such instances it is no charity to operate. The disease often remains dormant for years, or it extends inwardly, with little outward ulceration. The patient at last sinks gently into the grave from the constitutional depression of the disease, and not from a local drain. I would not operate when the patient is old and feeble, when the immediate effects of the operation is doubtful."

He gives us a case of which he says: " When the patient first came into the hospital the disease had advanced so far into the axilla that the removal of the whole of it was out of the question, and her pallid countenance, worn features, and general appearance of debility, seemed to preclude the idea of amputating such an enormous mass of active disease without extinguishing the vital flame. However, as her sojourn with us was prolonged, her general health improved ; but the disease advanced with rapid strides, till at last I could not, in charity, resist the desire to relieve her by one broad cut, believing,—and the event proved I was right,—that I could do it without endangering her life. One of our old physicians (I think it was Dr. FORDYCE) said, ' thank God for opium.' May we not indeed say thank God for chloroform in such cases as this ? The operation was performed without her knowledge, and she slept the same night better than she had done for months. She has now a healthy wound instead of a festering mass of corruption. This patient was forty-four years old, and had eight children. The disease commenced eight months after her last confinement, and eight months before her admission into the hospital, by pain in the axilla and shoulder : in the course of time a swelling was discovered in the breast, which gradually increased and became more painful. The mamma, though enlarged, was soft and elastic, very like chronic abscess. It ulcerated, and then the bleeding fungus mass sprouted and increased rapidly. The axillary glands were enlarged; nothing gave her any relief, and under these circumstances Mr. SOLLY thought it best to remove the entire diseased mass—not with the hope of curing her, but simply to relieve her from her agony and the profuse foul discharge.

"The progress of this case after the operation amply justifies the performance. I know the disease is not eradicated, but I hope it will take an inward course (strange idea), which is far less distressing to its victim, and the liver seems to be the seat of it.

"The chloride of bromine has been recommended as a specific in cancer (says the author). I have only tried it in two cases but in neither have I seen any beneficial results. In some cases I have thought that Dr. Arnott's apparatus has retarded the growth of this disease, and that in open cancer it has rendered the ulcer less offensive. On the whole, I have, however, been disappointed in its effects, and I seldom employ it now in any case."

In the performance of these operations the Doctor did not attempt to save much skin with the idea of getting the wound to heal rapidly by the first intention. "When the disease returns externally, it is generally on the skin which forms the edge of the cicatrix, which looks as if we might have prevented it if we had taken a little more of the integuments. You must also be very careful to dissect the pectoral muscle very cleanly. Do not leave any cellular tissue over it. In making the first incision remember that you make it below the mamma, and then the blood flows away from the knife, so that you see each tissue more distinctly as you expose it. Examine carefully the surface of the wound, in order to satisfy yourself that you have left no palpable piece behind. Also make a section of the tumor. This enables you to see if you have removed a circle of healthy cellular tissue beyond its margin. These cases, in private practice, are the most disagreeable and unsatisfactory that you can have to do with. Of course, your conduct will be guided alone by a strict sense of duty. You will neither be tempted to operate on account of the fee, nor to refuse on account of the discredit which follows in consequence of a return of the disease. You will calmly consider what is, on the whole, best for your patient, swayed neither by her fears nor those of her friends, nor by the prejudices of the practitioner in attendance, if such there be to contend with. It is most important that the patient herself should not know all your fears, and all your doubts. It is most important that one or more judicious friends should know your real opinion. In these melancholy cases the husband is seldom the person that can be trusted. Fortunately for us, as no man living can say positively that the disease must return, and must be fatal, we are justified in giving the sufferer the benefit of the doubt, particularly when hope itself may be the means of prolonging, if not of saving, life.—*Lancet*, January 15, 1856, p. 6.

Here is a fair, honest statement of facts and results, following the best practice known at that time. We see it gives but little hope of success. Nevertheless, it is a praiseworthy effort, and shows that the true philanthropist is ever ready and willing to embrace every opportunity, and give every available remedy a fair trial, that peradventure some good may be done, and our knowledge advanced. None but an envious bigot would object to progress in science.

DR. LANDOLFI'S METHOD OF TREATING CANCER.

BY DR. CHAS. LASEGUE.

Dr. LANDOLFI does not belong to the class of habitual advertisers of secret remedies. His method is not enveloped in any mystery. He is anxious to propagate his ideas for the sake of humanity, and submits them to the investigations of men of science. He has courted a publicity, which, enemies as well as friends, have admitted to be honorable, and therefore demands the attention he deserves. The principle upon which the treatment is based, consists in transforming a tumor of a malignant nature by conferring on it a character of benignity, which admits of cure. This transformation is effected by cauterizations with an agent looked upon as specific—the chloride of bromine, combined, or not, with other substances which have already been very frequently tried; but have hitherto been employed separately. The internal treatment is merely auxiliary. The formulas for the caustic are, except in a few cases, the following: Equal parts of the chlorides of bromine, zinc, gold and antimony, mixed with a sufficient quantity of flour to form a viscid paste. This is the formula the author chiefly used in Italy.

At Vienna he seems to have preferred a mixture of the same substances, in other proportions: chloride of bromine, three parts; chloride of zinc, two parts; chloride of antimony and gold, each, one part; made into a thick paste with powdered liquorice root. This preparation should be made in an open place, on account of the fumes which are disengaged. The essential element is the chloride of bromine, which, especially in the latter experiments, has often been employed without the addition of adjuvants: chloride of bromine, from $2\frac{1}{2}$ to 4 drachms; powdered liquorice, as much as sufficient.

According to Dr. LANDOLFI's views, the chloride of zinc is indispensable in ulcerated cancers, in which it acts as a hæmostatic. The chloride of gold is only rarely useful; it is particularly indicated in cases of encephaloid cancer, in which it exercises a special, if not a specific, action.

Cancers of the skin, epitheliomas, lupus, and small cysto-sarcomas, are treated with chloride of bromine, mixed with basilicon, in the proportion of one part to eight. At first he contented himself with spreading the paste on a cloth large enough to cover the diseased part, recommending that the thickness of the plaster should be proportioned to the depth to which it was intended its action should reach. He calculated that an epithem of a line in thickness should act to the depth of about half an inch. Subsequently he has had recourse to a more complicated method, and has adopted additional precautions, which we shall describe at length. The healthy parts surrounding the heterologous tumor are covered with strips of cloth, from an inch-and-a-half to two inches wide, smeared with a pomade, composed of four parts of chloroform and thirty of lard, or, what is better, of cold cream. The specific paste is afterward spread to the required thickness on compresses, and gently applied to the part affected. At this period of the operation the precautions mentioned above, in reference to the pharmaceutic manipulation, must be observed, and the patient must be kept near an open window, to avoid the injurious effects of the vapors of chlorine.

The paste is not to be spread on single compresses the size of the lesion, but on small portions of linen, placed side by side, or even imbricated, so as to insure closer contact with the subjacent parts. The application of the paste ought not to be extended to the healthy parts, its action being often propagated through a space of one or two lines. When the dressing has been so far completed, it is to be covered with a pleget of lint, and a layer of compresses retained *in situ* with strips of adhesive plaster. To a tolerably sharp sensation of heat, pains often very intense succeed, and last four or six hours, or even longer. A tablespoonful of the following mixture may be given every hour during the continuance of the pains: Hoffman's anodyne liquor, laudanum, of each six drachms; syrup of orange-peel, two ounces; distilled water, three ounces. The paste was formerly kept on from ten to fifteen days, but is now not allowed to remain on more than twenty hours. On removing the dressing, a line of demarcation is almost always found, separating the healthy from the morbid parts; the tumor itself is in part whitish, in part reddish, or marbled with yellow and blue. The caustic is replaced with poultices of crumbs of bread, or lettuce leaves, or with compresses smeared with basilicon ointment, which are renewed every third hour, till the eschar is detached. The pain progressively diminishes, if it has not completely disappeared, in proportion as the mortification advances. The line of demarcation becomes daily more evident. About the fourth or fifth day the cauterized portion begins to rise, and from the eighth to the fifteenth day it becomes detached, or can be removed without pain by means of a forceps, leaving exposed a suppurating surface, secreting pus of a good quality, and covered with healthy granulations. If any points remain of less satisfactory appearance, or still presenting traces of the former alteration, a little of the caustic paste is to be again applied. The wound is otherwise dressed according to the rules observed in the treatment of simple ulcers, whether with linen spread with cerate, or with balsamic ointments, or if the suppuration proceeds too slowly, with lint dipped in the following solution: Chloride of bromine, from twenty to thirty drops; Goulard's extract, from one to two drachms; distilled water, sixteen ounces. In the majority of cases healing takes place rapidly; cicatrization proceeds from the circumference to the centre; no complications supervene, and the cicatrix resembles that left by a cutting instrument. The general state is very satisfactory. He makes no change in the usual regimen of the patients. Those cancerous individuals for whom a perfect cure is not expected, experience remarkable relief. Notwithstanding the occasional great degree of local pain, febrile reaction is not demonstrable. All the observers who have witnessed the experiments, agree in their statements of the facts we have just described, from whatever point of view they may have regarded them, and they come before us with the guarantee of the most respectable authorities. As to internal treatment, we have said that the Neapolitan professor considers it as an auxiliary to which it is not always necessary to have recourse. He admits that the modification produced by the chloride of bromine employed externally, is not merely local, but that absorption of the specific by the skin or by the wound also takes place. It is as complementary to the treatment and to prevent relapses that he prescribes especially the internal preparation of the remedy, of which the following are the formula:

   R.—Brominii chloridi,   gtt. ij
   Pulv. semini fœniculi,   gr. xxiij,
   Ext. cicutæ,   gr. xij,   M.
   Et in pilulæ xx. div.

One to be taken daily for two months, and after that time two pills daily, (he does not say how long). Here follows another recipe:—

   R.—Chloride bromine,   1½ drops,
   Powdered seeds of water-fennel, grs. xv,
   Ext. hemlock, or aconite,   grs. viij. M.

Form pills, x.; one to be taken morning and evening for six months.

Lastly, in cancerous affections of the uterus, when the cancers of the parts most easily accessible are too extensive to be cauterized, when the cancerous cachexy has reached the highest pitch, he employs the following solution as a local modifier. Chloride of bromine from ten to twenty drops, distilled water, sixteen ounces. We have thus described most minutely the operative manipulation practiced by DR. LANDOLFI, and the pharmaceutic preparations to which he has recourse. Thus opening the way to a testing of his system, it would be of little use to have described those processes without at the same time endeavoring with the assistance of published observations briefly to estimate their value.

Our object, as we have already stated, is not here to enter into a discussion, the data for which would be wanting; this is a task we leave to others. But mistrust is so legitimate when its object is a specific medication, the least semblance of assent is so dangerous, that it is our duty before trying any of these methods, to make sure at least of the ground on which we stand. The first question is, that of the safety of the remedies. None of the observers, whatever may have been the amount of favor with which they regarded the system, who have closely watched trials made of it in Italy, or Germany, have noted any serious bad consequences as resulting from it. The local inflammation attending the elimination of the disease does not exceed the limits assigned it by the operator.

The general reaction is none, or is insignificant. All agree in stating that the patients were relieved, that they experienced no loss of appetite, or strength, or of sleep, but on the contrary, they from the first acquired a certain elasticity. The first datum, which appears to us to be established beyond dispute, is sufficient to save the conscience of the experimenters. It perfectly legitimizes the steps taken by the administrators of the several hospitals, which have furnished the professor with opportunities of propagating his mode of treatment.

The second question is more delicate. Were the tumors treated by Dr. LANDOLFI really of a cancerous nature? Were they not, to explain, a supposed success, errors of diagnosis? such as are too frequently made in putting forward a curative method? Never was there a period at which people were less disposed quietly to accept cancers diagnosed in haste, and to support a cure. If definitive opinions are not always uniform, all agree as to the necessity for close examination, and of not trusting to appearances. It seems to us more than probable that among the patients submitted to treatment, tumors and ulcerations of all kinds must have been confounded under too general a denomination. The descriptions are, by no means, all so explicit as to prevent us retaining some doubt as to the majority of the cases to which the most remarkable success is attributed.

Dr. LANDOLFI has, like all inventors, found, along with sceptics, partisans prone to enthusiasm, and consequently inclined to magnify the merits of the discovery by exaggerating the severity of the disease. But, whatever the narrow limits of our confidence, we willingly concur in the judicious opinions of Dr. CALDERINI. Carefully instituted experiments do not show that we have as yet obtained a specific for cancer. Those which have been made, justify us in believing that the plan recommended by Dr. LANDOLFI fulfills valuable indications; that it cures, without inflicting danger on the patients, tumors and ulcerations, the treatment of which was hitherto dangerous or difficult; that it furnishes the surgeon with a modifier of great power, as well as of perfect safety; that it improves sores for which we were hitherto deficient, even in palliatives; finally, that its author deserves to be distinguished from the crowd of inventors of whose discoveries nothing useful survives the day in which their panacea was proved to be neither specific nor infallible.

We have quoted faithfully what Dr. LASEGUE has said in his report of Dr. LANDOLFI's practice in the treatment of cancer. Notwithstanding the full and frank expression of opinion in any given case or disease, an *opinion* ever so strongly urged, is not worth one *fact* clearly demonstrated. Statistics would be, perhaps, less decisive, and we therefore bring forward some cases. Those published have been reported by medical men, who appear to be favorable to the system; but who rest it on scientific testimony. They have been too recently observed, to enable us to draw any legitimate conclusion as to the possibility of a relapse. Accordingly, we have abstained from speaking of the absolute curative rule of the treatment, convinced that it would be premature to sustain any conclusion of this kind, and to anticipate the future.

M. LANDOLFI's plan of treatment can only be judged by its actual results. (So of all other treatment of any and every disease to which the human family is heir).

Dr. DE BRUNN witnessed cures effected in the city of Gotha, whither Dr. Landolfi was summoned on the 13th November, 1853, to attend a princess of the reigning family, and where he treated in the space of two months about one hundred cancerous patients. Among the cases he reports, two especially deserve attention. The first was that of a woman aged 50, affected with a tumor of the breast, examined by Meckel of Berlin, and thus described by that microscopist: "The proper mass of the tumor consists of a reticulated structure, moderately supplied with blood-vessels in its fibrous tissue, in the midst of which meshes or alveoli, are plainly observed, filled with characteristic cancer cells. The tumor which was discovered about a year before, and at first lost in a general swelling of the breast, had become more and more isolated. It was hard to the touch, angular, had resisted the several means indicated, and had finally ulcerated, forming a serious excoriation with thick and elevated edges. The paste was applied on the 14th of November. On the 25th the eschar was removed with the forceps without pain or hemorrhage. The sore not being considered to be in a satisfactory state, was submitted to fresh applications of the caustic, which were continued until the fourth of December. On the 23d of January, the sore which had first been converted into a huge cavity, had cicatrized, with the exception of excoriations the size of a bean."

The second cure we shall quote, is that of a woman from Berlin, aged 60 years, who had for twenty years labored under a hard tumor, situated at the outer side of the left breast. This tumor which was during the last seven years in a state of ulceration, was very painful, had a bad smell, and gave rise to frequent hemorrhages. There was almost constantly a febrile condition, debility, and depression. At the time the treatment commenced, 23d November, the ulcer extended to the edge of the axilla. It was five inches in length, and three in width. The edges were callous and strongly adherent. Toward the posterior boundary was a crop of knotty tumors discharging sanies. Dr. LANDOLFI diagnosed a fungus hematodes. Microscopical examination by Meckel proved it to be a medullary fungus (a fungoid cancer). On the 3d of December the bottom of the sore was cleaned and covered with recent granulations. The edges remained hard and knotty. The application of the chloride was repeated. On the 15th, the granulations were well developed; the edges were inclined to close; a smooth and clean cicatrix was established. The patient's general state was improved. The improvement progressed rapidly, and on the 15th of January the cure was complete, after the disease had lasted for twenty years.

At Vienna Dr. LANDOLFI likewise treated, from the beginning of

June to the end of July, a great number of cancerous patients. The anonymous author who brought the results before the society of medicine, quotes at greater or less length, thirty-three cases which he observed himself, and which are thus analyzed: pseudoplasm, of the breast, seventeen females; of the nose, four; cancers of the lips, two; of other parts of the face, five; cancerous infiltration of the axillary glands, of the ribs and of the scapula, three; extensive carcinoma of the inguinal glands, one; encysted cancer, one.—*Dublin Quarterly Journal*, November 1855, p. 481.

With the amount of evidence above given, the acknowledged talent and high standing of the men who have given the above statements, can we for a moment hesitate to acknowledge that cancers have been cured without the knife? and if cured at all, the success is a victory over the knife, for by it very few, if any, are cured. Then why tie the hands of the profession, and blind the eyes of the community, and let them die of this direful disease, without the possible chance of relief? No philanthropic heart will allow such a feeling to dwell in his breast for a moment. The successful treatment of all diseases has been found out by the use of remedies untried till they were first used. Why then condemn the effort to find out a remedy for this dreadful scourge of the human race? Never let it be said that the energies of man shall cease, while the mineral and vegetable kingdom is yet so sparsely explored, and men of science are, with all the appliances of chemistry, daily bringing to light new discoveries, and new remedies, by the scientific analysis, and philosophical and chemical combinations of the elementary principles of the mineral and vegetable world. Rather encourage and aid by your efforts those who devote the best efforts of their life to making discoveries of these medicinal agents; and even if failure after failure should occur in their use, it will at least be the means that has aroused the profession to further investigation. Yet if you faint not, but battle on, seek for remedies, and face the monster disease, let him appear in what form he may, you receive the applause of your brethren; but when you touch the stereotyped treatment of a disease the remedies for which are acknowledged failures, you are cried down as an impostor, a quack, or a charlatan. Let this shame rest no longer on the escutcheon of our noble profession. The treatment of cancer is one of acknowledged difficulty. So is the treatment of Asiatic cholera. Nevertheless the physician does not refuse to seek for a remedy for that awful disease. So with scarlatina, a disease that has slain its millions. How many new modes of treatment are constantly handed forth to the profession, the reader of the medical journals can testify. Yet no one is called in question for that; but let him only make an effort to leave the old beaten track of treating cancer, and he is immediately cried down, and set at naught by the envious (only) of the profession. Let us be not weary in well doing, we shall reap, if we faint not, and thousands will rise up and call us blessed.

On the average duration of life in patients with scirrhous cancer of the breast. JAMES PAGET, ESQ., F. R. S., assistant surgeon to St. Bartholomew's Hospital, in a lecture delivered in 1852, at the Royal College of Surgeons, made a statement that the average duration of life in those cases, was thirteen months longer when the disease was left to itself than when it was removed by operation. More extended experience from the averages of a greater number of cases shows that this is not correct. We therefore insert a more nearly accurate statement of the average duration of life in both cases.

Mr. PAGET says, " Records which I have made, or collected of 139 cases of scirrhous cancer of the breast, watched to their conclusions, or to their survivals beyond the average duration, gives the following results:

" In seventy-five cases not submitted to operation, the average duration of life, after the patient's first observation of the disease, has been forty-eight months. In sixty-four cases submitted to operation, and surviving its immediate consequences, the corresponding average has been a little more than fifty-two months. The longest duration of life in the former class has been 26 months, in the latter class 146 months. The shortest, in the former, was seven months, in the latter, seven-and a-half."

Mr. Paget has given a table showing the length of time the patient lives, without an operation, and after the operation. He gives the preference to the operation by about ten per cent. Still, he dooms them all to a fatal end, sooner or later. If this proves anything, it proves the knife is not to be relied on for the radical cure of cancer. Why, then, should we not search for another remedy? With these remarks we close our quotations from European authors, though there are many more who might be noticed, who stand very high in their profession, and whose writings throw much valuable light on this subject. But the profession generally, especially those who feel interested on this subject, have the privilege of examining for themselves.

We have talent in America equal to any portion of the world—though our years number many less than theirs. We now come to notice some of our own writers on the subject of cancer.

And first, Dr. T. Gaillard Thomas, Professor of Obstetrics, etc. in the College of Physicians and Surgeons, New York. The position of Dr. Thomas as Professor of Obstetrics and Diseases of Women and Children, necessarily limits him to the diseases of women, etc. He therefore confines himself on this subject to cancer of the uterus, which he gives in the following table, under the head of "*Malignant Diseases of the Uterus.* First.—Cancer; encephaloid; colloid; scirrhus; cancroid; fibro-plastic; recurrent fibroid; epithelioma; corroding ulcer; cauliflower excrescence."

We shall only notice what he has said on cancer, and refer the reader to his valuable work for his description and treatment of the other diseases here named.

"*Definition and Synonyms of Cancer of the Uterus.*—This disease, which has been described under the synonymous terms of carcinoma-uteri, and ulcerated carcinoma, may be defined as a degeneration of the interstitial tissue of the uterus, characterized by grave constitutional implication, great tendency to molecular death, and a certainty of reproduction if removed by surgical means."

The Doctor here makes a frank acknowledgment that the remedy for this disease is not to be found in a surgical operation. Then, as philanthropists and guardians of the health and safety of the human race, as far as human means can be employed, why shall we not seek after, and cease not to try other remedies, until peradventure we may find the right one?

According to ROKITANSKY, the following average scale may be adopted as representing the preference of cancer for various organs. First, the uterus, the female breast, the stomach, the large intestines, and especially the rectum. Next comes cancer of the lymphatic glands, etc. The great frequency with which the uterus is thus affected, may be judged of by the statements of Prof. SIMPSON, based upon reports made under the registration act, during a period of five years (from 1838 to 1842) for England, exclusive of London. The number of women who died of cancer was 8,746; of this number 3,000 died of cancer of the uterus.

These statistics prove that cancer is nearly three times more frequent in women than in men, and more than three times more frequent within the uterus, than in any other organ of the female. M. BECQUEREL asserts, that, in spite of its great frequency, cancer of the uterus is not a disease of which the history has been long known. Dr. THOMAS says, that it was not understood as we understand it to-day, is most true; but the ancients surely had a great deal of very accurate knowledge concerning it. HIPOCRATES, de morbis Muliorum, describes it at length, declaring it to be incurable; ARCHIGENES wrote an able chapter upon it, describing the ulcerated and non-ulcerated forms, and the peculiarities of the discharges. His article is preserved by AETIUS, who entitles it, " De Cancris Uteri." The Arabians likewise were familiar with it; they allude to it in such a way as to satisfy us that they understood its nature and dire effects. Upon the revival of Gynæcology in France, the disease was confounded with fibrous tumors and parenchymatous inflammation, or rather with its resulting hypertrophy. ASTRUC described " Scirrhus," as the result of abortion in 1766, and the confusion which attached to its description, extended long after him. It characterized the times of RECAMIER and LESFRANC, and even so late as our own period we see the view endorsed by Drs. ASHWELL, MONTGOMERY, DUPARCQUE, and many others.

Messrs. BLATIN and NIVET, in expressing the belief that scirrhus results from chronic inflammation of the parenchyma, append the following foot note: Paul of Ægina, Galen, Andral, Broussais, Breschet and Ferrus, Piorry, Bouillard, etc., place scirrhus, among the terminations of chronic inflammation. Some of them, however, admit the existence of a predisposition.

Dr. THOMAS says: " Cancer of the uterus bears no resemblance to fibrous tumors, polypi, or parenchymatous engorgements. It arises from a constitutional vice, and is never the result of chronic inflammation or any other purely local condition, (*very true*.) It is incurable, and if removed by surgical means, will invariably return." Then shall we succumb to our inveterate foe? No! let us seek after the truth, lay aside old opinions, truly valuable as they were in their day, if for no other reason, than to prompt us to further research, that by unwearied effort we may ultimately find the truth, if only by little and little. Still preserve what we gain, and press on for further developments by the aid of unfailing effort, till the truth in all its plainness shines forth upon our intellectual vision, and those obscure and only partially manifested truths, take the place of error and mistaken theories, and we come to see the true nature of cancer, classify it according to the basement membrane on which it originates, analyze and ascertain the true chemical nature of its secretions by extracting from its cells the appropriate juices, and thereby be enabled to apply the remedy that will destroy these morbid and vitiated juices, so arrest the progress of the disease, and cure the patient. This never can be done by the knife, which is abundantly proven by all celebrated surgeons, who say you may cut it out, but it will return. Then why, O why stick with so much tenacity to a remedy which you know and acknowledge will not cure?

There is however much valuable information in Dr. THOMAS' work, which we cannot further pursue at this time, but which the inquirer after truth may find by perusing his pages on this, as well as many other subjects. The distinctions he makes between cancer, encephaloid, colloid, scirrhus, cancroid, fibro-plastic, recurrent fibroid epithelioma, corroding ulcer, and cauliflower excresence, may all be readily classed, by first ascertaining if they possess the true cancer-cell, then place them as they should be classed, according to the basement membrane from which they start, and you have no difficulty in your diagnosis. They are all cancer, if they have the true cancer-cell—if not, they are not cancer, but something else to be diagnosed according to the true character developed. Simplicity and plainness lead to the truth, while multiplicity in complication in similar circumstances lead to confusion, doubt, and uncertainty, and of course divided effort to obtain the same end. Have but one name for the same disease; let it be cancer, but classify according to the primary location. We then look as it were to the bottom of the matter; we select the means to detect them and arrest the intruder, be his location where it may, if it be in striking distance of us. When it is located on the basement membrane of an internal organ, as the stomach, intestines, kidneys, liver, or pancreatic gland, where we cannot get at it, we acknowledge we know no remedy as yet that will reach it. Only palliatives then can be used.

S. D. GROSS, M. D., Professor of Surgery, etc., Philadelphia, writes as follows:—

LEPOID.—The lepoid formation is most generally observed upon the face, nose, and forehead of elderly persons, usually males, of a delicate, florid complexion, with a habitual tendency to the capillary vessels, light eyes, and light brown or reddish hair. Although occasionally single, it is more commonly multiple. It generally makes its appearance in the form of a small circumscribed speck, not larger perhaps than a mustard seed, and of a dirty grayish color, which becomes covered with a very rough brownish crust or scale. This, falling off, is soon succeeded by another of the same complexion, form and consistence. Thus the disease is often kept up for many successive years. At last, however, ulceration sets in, and the dermis exhibits a red glossy surface, spicular, pitted, or granular, and throwing out a thin, elaborated pus. The skin, upon inspection, is found to be almost of a gristly hardness, its internal surface being studded with numerous little whitish, rounded bodies, connected together by a dense grayish substance. The progress of this disease is attended with hardly any pain; but the patient is generally very much annoyed with itching, leading to an irresistible desire to scratch, which always aggravates it. The nature of lepoid is undetermined. Without being able to speak positively, I am strongly inclined to believe that it is merely a variety of lupus, or epithelioma, a supposition deriving plausibility from the circumstance, that although generally disposed to remain long stationary, or to make but little progress, it often ultimately takes on malignant action, pursues the very same course as the milder forms of lupus. Here the doctor brings the disease ultimately to so resemble a lupus, as to call it an epithelioma; and what is an epithelioma but the mildest form of cancer, designated and properly named from the basement membrane on which it starts? Nevertheless, when these "roundish, whitish bodies," which the doctor so clearly describes, are closely examined, they will be found to contain more or less of the cancer cells, which Dr. BENNET so clearly describes. According to our position, if the embryo does not contain in miniature the characteristics of its true nature, it never will by any process of morbid action take on a new form of disease, so as to change its character. And although it may require many years for the disease to clearly manifest its malignity, yet the seeds of that malignity were sown in its most incipient stage, and like the germs of other diseases, it requires time for its development. Lepoid, being barely located on the skin, and imperfectly covered by the cuticle, and being exposed, from its location, to the constant action of the atmosphere, holds it in bay for many years. Nevertheless, the lion is there, with all his ferocious nature, and only requires time to show his true character.

By examining plate No. 191, in Dr. GROSS' Surgery, vol. I, p. 634, you will there see distinctly laid down those cells that now are looked upon as the only sure evidences of this formidable disease.

The Doctor says, "the best remedy for this disease is non-interference." The only reason that I can find for such a statement from such a gigantic mind is, that he is wedded to the knife, and he sees and knows the knife will not do. Let inferior minds suggest another course of treatment.

The sons of Æsculapius never yield to their enemy till by force of power they fall in the battle-field. If we fail with one, we up and try another weapon, until at last we find the blade that cuts the monster down. If the surgeon's knife fails, we resort to the chemist, the pharmaceutist, and the combination of remedies, until we find the remedy that conquers the lurking monster and routs him from his secret lair.

LUPUS.—Under the term lupus, Dr. GROSS comprehends two varieties of the disease, "the chief peculiarity of which is a tendency to destructive ulceration of the skin and alveolar tissues, or of those and the more deep seated structures." He says, in point of fact, the two affections are identical. The only difference between them being that the one is milder than the other, more tardy in its progress, and less disposed to spread; its ravages being generally limited to the cutaneous textures, or to the parts in which it is originally located. The former of these he calls the non-exedent, stationary or serpigenous ulcer. The latter, the exedent, eating or corroding ulcer. It is also known as the cancroid or voracious ulcer.

Could any man describe more clearly the fungoid and fibroid varieties of cancer than this master pen has done it? They commence a little differently, but ultimately prove to be malignant. The difference is to be found in their location on different basement membranes. They are both cancer. Then why not place them in their proper classification, and treat them accordingly?

Dr. GROSS says, "in the treatment of this variety of lupus, none but the mildest and most soothing applications should be employed." This is tantamount to saying, let it alone. The less you do for it the longer the patient will live. Nay, my dear friend, let us take the lion by the beard while he is yet a whelp, and destroy his life, and remove all his strongholds.

In the meantime, as the Doctor truly says, "special attention must be paid to the state of the general health." This every physician will certainly attend to. In the second form or variety of lupus, the excedent, corroding or devouring lupus, he frankly states it is "epithelioma." What is epithelioma but cancer of the outer basement membrane? If so, should we hesitate to treat it accordingly? The Doctor says, " the knife, or some suitable escharotic, as the acid nitrate of mercury, or the Vienna paste." Here you are right. But your escharotics are not powerful enough. Turn back to the remedies prescribed in these pages for that class of cancer, and you will find the remedy. I find myself very unwilling to differ with a man for whom I have the highest regards; but sad experience in myself, and much experience with others, compels me to do so.

For many years I have almost ceased to use the knife, hence my want of acquaintance with the late surgical writers. This will give the reason for the remarks I here make on ERICHSEN's description and practice on this disease. This celebrated surgeon has given the best description of the different varieties of cancer, of any author I have seen; indeed he has given a good classification of the disease. Those who may read his classification and my classification, would be likely to draw the conclusion that I had founded my classification on his. But this is not the fact. I never saw ERICHSEN's work on surgery, until this day—the present day of this writing, after all I have heretofore said on this subject, had been written. Indeed, I may have quoted him as quoted by others, but not from any examination of his work. Here let me say, that his description of cancer, and he makes all the varieties cancer, is so clear and correct that it should not be ignored by any. His treatment lacks specification and special directions, but the aim and intention is good. His failures will be found, however, to follow the knife much more frequent than they do the caustics. The day is coming when the knife will be ignored, and the treatment of this disease will be taken from the hands of the quack and pretender, and treated according to the true principles of physiology and pathology, with a correct therapeutical knowledge of the power of remedies used for the cure of this disease.

CANCER ON INTERNAL ORGANS.

The following case is well worthy of being kept upon record. This case terminated fatally at St. Bartholomew's Hospital. It is a well-marked instance of infiltrated scirrhus of the esophagus, arising suddenly without assignable cause, yet suspected by Dr. FARRE before the death of the patient. This patient was fifty-five years of age, a plasterer by trade, admitted into the hospital on the 19th of September with symptoms of general ill-health, and complaining more particularly of excessive dysphagia, or inability to swallow. He stated that up to five months ago he was quite well; he might have been taken indeed as a model of good health; "he never knew what a day's sickness was," and would now be quite well but was starved on account of his loss of food, caused by the excessive pain in swallowing. Scattered over his skin were several raised malicious looking spots, which first roused the suspicion of Dr. FARRE and Mr. LLOYD that the disease was of a cancerous kind. A probang was passed into the esophagus, but was stopped by a very evident narrowing of the canal, and the use of the instrument was followed by hemorrhage, so it did not seem advisable to repeat the operation. The patient had the usual nutritious food prescribed, with milk and mild dumulcents, but after much suffering gradually sank on the 25th. The post-mortem examination, carefully conducted by Mr. CALLENDER, was particularly interesting. The friends of the deceased were extremely anxious for a full examination, as they could not understand the suddenness of the illness. The man's son attended the necropsy. " The lower third of the esophagus as it joins the cardiac orifice of the stomach, was almost entirely obliterated by a mass of infiltrated scirrhus, or cancerous disease. The lungs, though the man was stated to have been previously in perfect health, were filled at several points with healed or obsolescent tubercles; the liver, also, was filled with a similar deposit. The small intestines, at one or two spots, were also obliterated by cancerous growths, leaving the canal of the gut not much larger than a common quill." This form of the disease in the esophagus is rare, though a similar case occurred in St. Bartholomew's in 1851. The singularity of this case deserves a record. This disease "began with pain of a severe kind about the shoulder and lumbar region, ending in the most excessive difficulty in swallowing, which was found after death to depend on cancerous degeneration of the esophagus."

A case of marked encephaloid is noted by Dr. HANDFIELD JONES, which had proceeded to a considerable extent without producing any marked symptoms, as well as another instance of a man who had died of peritonitis, in which, at the post-mortem examination, a thick encephaloid mass, ulcerated on the surface, just above the cardiac orifice, was accidentally found extending for an inch and a half.

In the *Lancet*, 1832, are two very interesting cases of disease of the esophagus, in which the affection had assumed the character of softening, subsequently dwelt upon by ROKITANSKY. (See Guy's Hospital Reports, vol. vii., p. 141). A case of typhus is given, in which the esophagus was found entirely softened. These cases, and others which might be cited, corroborate the suddenness with which the results of malignant disease make their appearance in this part, which is, as a general rule, free from any other morbid affection. These cases may put us on our guard in giving a diagnosis in diseases located in these organs.

Diagnosis of Tumors of the Breast.—" What are the symptoms by which a cancer of the breast is to be made out? Cancer is essentially of a progressive character; it never remains long in one tissue, but sooner or later involves all with which it comes in contact, and it is through this pathological peculiarity that it produces symptoms in the skin over it and parts beneath it, of great diagnostic importance." (THOMAS BRYANT.)

No better description and satisfactory definition can be given of this dire disease, than that which Dr. BRYANT has given. We therefore shall quote largely from him.

How Cancer affects the Skin.—"The skin is affected slowly, but in a very characteristic way. It first becomes more or less fixed to the growth, and, instead of rolling freely over the surface, as it invariably does in a simple tumor, it seems tied or drawn down to the parts beneath. A dimpling of the skin next makes its appearance,—this symptom passing into a complete puckering,—at which stage of the disease the integument will have become completely fixed and immovable. As the disease progresses, the skin becomes infiltrated with the cancerous products, and is made a part of the diseased mass. As this progresses the so-called ulceration of the integument may make its appearance. This ulcerating surface always assumes a characteristic aspect. In some cases the skin becomes infiltrated with isolated tubercles, varying in size from that of a millet seed to tubercles of larger dimensions. When any of these conditions of skin, with a tumor infiltrating the mammary gland, wholly or in part, and more particularly the latter or tubercular form of infiltration appears, the cancerous nature of the disease is tolerably clear." The mobility or immobility of the tumor is of great importance in deciding the character of the disease. As cancer affects the skin by infiltration, so, practically, it affects the parts beneath, and, as a consequence, we find symptoms produced of great value in a diagnostic point of view; for a "cancerous tumor, instead of moving freely over the pectoral muscle, as it does in innocent affections, and in health, becomes, as the disease progresses, somewhat fixed, and at last immovable." This is in the course of progress, but not so at first, "the breast gland, pectoral muscle, and probably skin, all becoming, by general infiltration, involved in the same destructive influence at the last stage of the disease." The infiltration of the absorbent glands must not be overlooked. Should you find enlargement of the axillary or clavicular glands associated with a doubtful tumor of the breast, you may regard such symptoms as strongly indicative of the disease being cancer; for cancer rarely exists any length of time without causing such symptoms.

You may find glandular enlargements associated with inflammatory affections of the mamma; but, under such circumstances, the axillary glands are usually more tender on manipulation than they are in cancer, and the enlargement is more rapid,—the cancerous infiltration, as a rule, appearing in its early stage as a painless chronic infiltration. There is still another symptom which is of great value in determining the nature of a tumor of the breast, for it is rarely present in any other affection of that organ than the cancerous, that is, neuralgic pains down the arms. It indicates the strong probability that the deep cervical or axillary glands are infiltrated with the disease, and are pressing on the nerves, causing pain, although these glands may not be felt. Indeed, neuralgic pains in any part of the body, associated with cancerous disease, or even after its removal, should be always regarded with suspicion, for they are often indicative of secondary cancerous deposits having taken place, either in the nervous centres, or along the nervous course. Retraction of the nipple is not an invariable indication of cancer of the breast. There are many other conditions of the breast in which the nipple may be retracted, and yet the breast may not be cancerous; so it is with a discharge from the nipple. In all cases of general diseases of the breast gland,—inflammatory, cystic, or cancerous,—there may be a discharge from the nipple. In the inflammatory, it may be watery or purulent; in the cancerous, it is generally sanguineous, and little in quantity; in the true cystic, it is viscid, more or less blood-stained, but profuse. It does not exist in all cases. As a means of diagnosis in the true cystic disease, it is of great importance.

Cachexia.—Dr. BRYANT says, "I have no belief in the existence of a special cancerous cachexia." It may be present in cancer as in any exhausting and wasting disease, but that of cancer differs in no essential point from that of any other disease. Cachexia always indicates the existence of a disease that is undermining the patient's strength.

We are desirous of giving all the remedies of any celebrity, that the profession may make such selections ás in their judgment the case requires. I am not wedded to any stereotyped practice, though in the general, I prefer the one I have given; yet, others may prefer some other practice. In order, therefore, that a fair trial may be given, and choice be made, I give the claims laid to other remedies.

We have said what we believe to be true in relation to the knife. We are still of the opinion that it is too uncertain a remedy to be relied on, and should rarely be used. Dr. RICHARDSON used a styptic colloid.

RECIPE FOR MAKING DR. RICHARDSON'S STYPTIC COLLODION.

R.—The fluid consists of ether and alcohol, the ether being in excess, saturated with tannin and gun-cotton. The fluids used as solvents must both be absolute, and the ether should have a low specific gravity, the boiling point not higher than 92° or 94°. The fluid diluted in equal parts with ether, may be used in the form of spray; but the most common way of applying it is with a brush, precisely as gum is applied.

Acetic acid has been alternately taken up and laid aside by the profession in the treatment of cancer. Dr. J. HUGHES BENNETT, in his work on Cancerous and Cancroid Growths, published in 1849, gives us the following passage, which shows that he was one of the first, if not the very first, to point atttention to the fact of cancer being acted upon by acetic acid and other agents. He says: " We have seen that certain chemical agents have a marked effect upon the cancer-cell. Acetic acid especially dissolves the cell-wall more or less, and strong potash reduces the whole to a granular mass. The continued application of these agents, therefore, would tend to dissolve the growth, if it could be brought into direct contact with the cells, and need not necessarily excite such irritation as to cause fresh exudation. The only objection is, the utter impossibility of affecting the whole mass, even in cases of ulceration, and preventing the formation of deep-seated cells, while the superficial ones are destroyed. In certain cancroid growths, especially epithelial ones, the application of acetic acid is an established remedy, and should always be tried whenever it is thought possible to bring the fluid successfully in contact with the entire mass of the disease."—*British Medical Journal*, *Nov.* 24, 1866, p. 593.

Further Reports of the Effects of Acetic Acid in the Treatment of Cancer by Dr. JAMES MORTON, *communicated to the Glasgow Medico-Chirurgical Society.—Several Cases of Cancer treated by* Dr. BRODBENT's *Plan, Med. Journal,* 1867.—Dr. M., in his remarks on these cases says: "Perhaps in not one of these cases has acetic acid, as thus applied, proved adequate to the complete cure of a single sore, or the eradication of one cancerous growth; and yet from its application some benefit may have been experienced by the sufferers. The most remarkable circumstance is the comparative freedom from pain when diluted acetic acid was used, which contained about four per cent. of real acid. Considerable pain was complained of when the glacial acid was employed, but this contains as much as eighty-four per cent. of real acid; and the pain produced by caustics has always been one of the chief objections to their use. In no case has there been any apparent acceleration of progress. Under the fear of such a result, the idea of making punctures into malignant growths has been long discountenanced, and can only be justified with a view to their amelioration, not removal. A more or less distinct retrogression in respect to size, seems to be one of the effects of the use of the acid, and a corresponding retardation of progress." "We are not yet in a position which entitles us to say that acetic acid eradicates cancer, or to vaunt it, in common language, as a cure for cancer." "Who that knows anything of this painful malady does not ardently wish for such a cure? Hundreds of such have been handed from father to son, usually as family secrets, but their composition not being revealed, the profession regards them with distrust. For this reason the profession has been accused of apathy and indifference in regard to such matters. No accusation could be more utterly false. The profession is thirsting for such a remedy; but it must deserve the name, and must not end, as heretofore, in disappointment." Shall a failure in any one remedy put a stop to effort by the profession in trying to find one generally successful? You might with equal propriety say, because a physician loses several cases of a malignant fever, he should not try to find a better remedy. No: we will never give up the effort till some remedy is found. Though like BREE you may destroy a hatful of eyes before one be saved, yet he persevered and finally couched successfully for cataract. We *must* have a remedy, and one that does not end in disappointment. You must have charge of the case while it is medicable, and not after the constitution is broken down by the disease. Deal fairly, ye opponents of remedies. Dr. MORTON goes on to say, "The ready welcome that has been given to the recent proposal of Mr. BRODBENT, is a proof that professional men are open to the consideration and trial of any plausible suggestion for the cure, or even the relief, of such a dire disease. So marked has this been, that here, in Glasgow, we have about as much experience of its effects and mode of action as can be found anywhere at present; and although the results have not been all that could be desired, yet ameliorations have taken place, apparent retardations of progress, and relief or mitigation of suffering. A conviction has been forced upon me," says Dr. MORTON, "that the remedy possesses a certain degree of power over cancerous formations. It is also possible that our present modes of application do not give the remedy fair play. I suspect we are too sparing of it, and it is not unlikely that some benefit would be derived from its hypodermic injection into the adjacent tissues of such tumors, as well as into the growth itself. The instrument being too small, a larger one might be more efficient and equally safe." This, I know from experience, to be a good suggestion. Many objections were raised to this remedy by Drs. ALEXANDER SIMPSON, MCGREGOR, ADAMS, ANDERSON, and others, who were present at the Society when Dr. MORTON read his paper. Many of them objected to the syringe, and said they could not pass the fluid into the tumor with such an instrument. This objection could have been obviated by a different syringe and a different mode of using it. When I inject a cancer, I first drill the thick crust on the surface, then with a crooked, spear-pointed bistoury, I pass it through the crust and lightly into the tumor below (this part of the tumor I have always found to be spongy); I then use a gutta percha syringe, with a point an inch long, inserted into the orifice made by the bistoury to the bottom of the puncture and force the fluid into the tumor, or rather below it. It will spread all under the tumor, or if the tumor is so large that one injection will not reach all its base, I make another until all the base of the tumor is reached. It will bleed but little; the injection, if of proper material, will check all hemorrhage. After this, poultice it, until the whole lump drops out. This will take place in 6 or 8 days, and the inner surface of the wound will be perfectly smooth if all the cancerous matter is removed—if not, you will be able to detect the cells that remain, and you must remove them by the application as prescribed for that purpose. We will now close what we have to say on the treatment of cancer at present, by showing the different effects of

CITRIC, ACETIC, AND CARBOLIC ACIDS IN CANCER.

We are indebted to Dr. JOHN BARCLAY, for the facts contained in this part of our work. He says a patient having a large and excessively painful cancerous tumor of the neck, behind the angle of the jaw, which from its size, situation and extent of its attachments, held out no hope of its successful removal by the knife, requested me to allow her to make a trial of citric acid (having seen it noticed in a newspaper), and her clergyman had also strongly recommended it. He at once assented, thinking it could have little effect either for good or for harm. But when he called in a few days after, he was somewhat surprised, that since she had applied the lotion, composed of a drachm and a half of the acid, dissolved in eight ounces of water, she had almost no pain, in the tumor. This he was the more astonished to hear, as nothing of the anodyne class seemed to have afforded the slightest relief from pain before, with the exception of the hypodermic injection of morphia, and even this, she said, did not remove the pain as effectually as the citric acid lotion: and besides, the relief from the latter was much more permanent. This treatment was continued for some weeks, with the effect that the patient improved considerably in looks, health, and spirits.

To show that this was really due to the change of remedy, he ordered her to discontinue the wash for a week, during which period of omission the pain returned with as great severity as before, compelling the patient to resume the application, which again brought relief. It now occurred to him, remembering the solvent power of acetic acid over cancer cell-walls, to try what effect that acid would have when applied to the tumor. By this time the skin had begun to give way, and a sore to develop itself on the surface. He then ordered vinegar to be applied, and was glad to find that this controlled the pain, quite as well as the citric acid had done before.

Mr. MASON, the Senior Surgeon of Chalmers Hospital in that place, who had used the citric acid lotion with equally beneficial results in another case of cancer of the breast, totally unsuited for excision, from its extent, adjacent glandular enlargement, and the length of time it had existed, now changed it for the vinegar applications, and he had no reason to be disappointed. He found that its anodynic power was equal to that of acetic acid, and also that it possessed advantages which the acid formerly used by us possessed in a much less degree. In the case of the wounds in the breast, which were numerous, and all partaking of the character of cancerous sores, in a most marked degree, it was noticed that after the application of the citric acid the thick, serrated and everted edges did not seem quite so thick, serrated and everted as before, but looked thinner, softer and with less induration around them. The application of the acetic acid lotion, as he hoped would be the case when gradually increased in strength, from that of common vinegar upward, produced a much more marked effect, for the edges began to thin down much more rapidly, granulations of a seeming healthy character arose in the centre of the wound, all fetor disappeared from the discharge, and even attempts at cicatrization began to take place, sufficient in several of the smaller ulcers to close them in altogether. The acetic acid in varying degrees of strength had now been used for a month in both cases. And looking back over this period, and over the month during which the citric acid had been used, to the condition of both patients, previous to the use of these remedies, they were of opinion that not only had their states of health, and the appearance of the disease in each, not become worse, but that both had sensibly improved.

Both patients ate and slept much better than before, and were able to go about their usual household occupations with ease and comfort. The tumor in the neck had become decidedly less. There were attempts at skin-forming process at several points on the edges of the sore, and pain in it was reduced to a minimum. As for the case in the breast, the woman had so much improved in health that it would have been difficult to recognize her as the same thin, cachectic looking creature of three months back. The cachectic look had wholly disappeared, and no one by looking at her now could have supposed she labored under a disease of such a serious nature, and which had progressed so far. The tumor itself was no larger than it was three months before, perhaps rather smaller, and several of the smaller sores had healed over entirely. And now it was resolved to try the effect of carbolic acid in the above cases. This was commenced on Dec. 8th, of the year previous to this history.

About this time two other cases of cancer applied for advice. One was an extensive tumor of the neck of the uterus, and implicating the whole of the vagina, accompanied by very great pain, and a most profuse and exceedingly offensive discharge. So fetid was the discharge that no one could stay, even for a short time, in the room with the patient. The other was an enormous scirrhus tumor of the breast, of very rapid growth. It had been in existence only four months, and already it had extended from the floor of the armpit almost to the sternum. The subject of it had been in one of the largest hospitals, and had received no relief, either there or anywhere else, from the extreme pains and the horrible fetor of the discharge. The carbolic acid in the form of a very dilute lotion was used in all four cases, with the following results:—

In the case of the tumor behind the jaw, the lotion was about as effectual as either the citric acid or acetic acid lotions. Applied in this weak state, its solvent effect was much the same as that of the citric acid; but applied in a more concentrated form the effect was a most vigorous eating away of the tumor, and with much greater rapidity than by the two acids formerly used. There were feeble attempts in skinning under the use of the dilute carbolic acid, in this case, and none of course, when the stronger solution was employed. In the case of the mammary tumor, which had been treated before by the citric and acetic acids, the report was that the pain was as effectually controlled by the carbolic as by either of the other acids. That under its use, the thick serrated edges disappeared much more rapidly than with the other two; and that when the weak solution was employed cicatrization was seen going on over many of the sores, whenever the cancerous excrescences were eaten down to below the level of the surrounding skin. In the case of the disease of the uterus and vagina, the effect was equally striking. Whenever the weak solution was employed, the pains almost entirely disappeared, and with them the horribly offensive discharge. The poor woman from wishing herself dead began to have her spirits raised, and to eat and sleep well. Now no fetor was perceived by those in the room with her, and the general improvement in her appearance was beginning to be visible, when a severe attack of hemorrhage nearly carried her off, since which time her progress towards recovery has been very slight. A like result was obtained in the fourth case, that of the large mammary tumor. The pain instantly on the application of the carbolic acid lotion, disappeared, as if by magic; and the fetor of the discharge was very much lessened. The tumor was extirpated a few days after the application of the acid, and the case has gone on well since. I may mention, that in treating cancer cells under the microscope with acetic and carbolic acids, in varying degrees of dilution, he found that in about equal strength the carbolic acid dissolved the cells much more rapidly and effectually than the acetic acid, and caused the nucleus also to disappear almost entirely, when applied in a concentrated state. From the above experiments, it appears that the three acids have about the same effect in removing pain from cancerous growths, and of correcting fetor from cancerous discharges. They all have solvent effects on cancerous tissues, but in my hands they never have cured a cancer.

    R. Acidi carbolici,   ʒiss to ʒij.
      Spiritus vini rectificati, ʒj.
      Aquæ Destil.,   OOij. M.
    Apply on a wet cloth.

PROFESSOR SALISBURY is of the opinion, that "as long as the parent gland cells organize normal products in the normal quantity, no such thing as diseased action could possibly occur." As soon however as these normal conditions are disturbed, the physiological balance is deranged, and pathological conditions arise, at first functional, but which continued long enough, tend to organic changes. The human family is more or less constantly exposed to conditions which have a tendency to disturb the physiological functions. We have but little control over the cell elements and the filamentous tissues, formed by the metamorphoses of these cells—after they have been organized, and have escaped from under the influence of the parent gland cells. "It is upon the parent gland cells that all healthy and diseased impressions are made; and it is to these cells that we must look for diseased states, and to these we must address our remedial agents, if we expect to remove successfully the cause of pathological derangements; if we cannot correct their morbid actions we must remove them. When a case comes to the physician, very frequently the primary causes of disease have ceased, as in the case with cancer, and the enemy is passing out to other neighboring cells. He has now to do with the resulting disturbance, which to all intents is the cause, and becomes the enemy he has to combat. But so far as the primary cause exists in force, it should be first removed, and then the results of the diseased states of the gland cells sought out, and means resorted to, which will in the most direct manner restore the healthy functions. We must not forget, that each parent gland cell has all the elements of an independent organism, possessing a vitality independent of that of the systematic life. Each cell feeds and digests its food, assimilates it, grows, organizes, and eliminates. In the normal condition it organizes those cells and filamentous and other products that are needed for the physiological issues. When its food is imperfect, its processes of digestion, assimilation, and organization become deranged, and pathological products are the result. These pathological products vary with the varying conditions." Under certain diseased states lithic acid, and the lithates are formed in too great quantities to be kept in solution by the eliminated fluids, and hence they accumulate in certain tissues. Under other pathological conditions oxalate of lime is formed. This body for the most part being insoluble in the fluids eliminated, it accumulates first in the tissues in which it is formed; and later, if the diseased states continue it extends to the other parent gland cells and tissues; and finally the entire organism becomes loaded with this salt. In still other pathological states, cystine, another insoluble body is formed by the parent gland-cells, and this body, like oxalate of lime, not being soluble in the fluids eliminated, gradually accumulates in the various gland-cells and tissues.

Under other abnormal states, the phosphates of alkalies and alkaline earths, are formed, and gradually accumulate as in the case of cystine and oxalate of lime. Under still other conditions, all of the pathological states previously mentioned may be present, under which circumstances we have the two, more, or all of the specific states united. Now these are precisely the conditions that develop the cases of cancer, and to get rid of their dire ravages, as they are local in their origin, let us use those remedies that have an affinity for, and effectually destroy every cell that has been placed under their influence. Then and then only, are you sure of eradicating the disease.

CUNDURANGO.

NOTE BY THE PUBLISHER.

As these pages are passing through the press, a new remedy has been proposed for cancer, which, if all is true that is claimed for it, the great desideratum pointed to by the author of this book—the discovery of an agent that will reach the specific virus of cancer, and neutralize it—is already met. We give the facts that have come to light on the subject for what they are worth, premising, however, that the observations already made are too recent to enable us to speak with confidence as to its claims as a remedy for this terrible malady.

Cundurango is a shrub which, it is said, grows in some "almost inaccessible region" in the Andes, in Ecuador, South America. Its virtues were discovered originally in a very remarkable way—a woman, in attempting to poison her husband, cured him of cancer instead! On this hint certain physicians of Quito experimented satisfactorily with the new remedy, the United States Minister to Ecuador obtained some of it and sent it to the State Department at Washington, whence it was distributed to medical officers of the army and navy and others, to experiment with it. The observations —we can hardly, as yet, say experience—of the army and navy officers were unfavorable; but one physician, Dr. D. W. BLISS, of Washington, is quite favorably impressed with it, even to the extent of believing it to be "quite as reliable as a specific for cancer, scrofula and other blood diseases, as cinchona and its alkaloids have proved to be in zymotic diseases."

Dr. BLISS has made a record of his experience in the pages of the New York *Medical Journal.* He says:—

"Mrs. Matthews, the mother of Hon. Schuyler Colfax, had been the victim of mammary cancer for a long period, which had already assumed secondary and constitutional symptoms in a marked degree.

On the 29th of April last, I placed her on the decoction of cundurango, and had the gratification of observing an early and decided change for the better, in both the local and general conditions. One of its almost immediate effects was the relief of pain, and a free diaphoresis, characterized by an odor distinctly observable of the infusion itself. Upon the return of Mrs. Matthews to her place of residence in Indiana, I still continued to direct her treatment, and furnished the requisite supplies of the medicine.

"On the 9th of May, just thirteen days after the commencement of the new remedy, her husband addressed me a letter, from which I make the following extracts:—

*   *   *   "The stony condition of the tumor has given place to softness. This morning I notice about one-third of the surface has turned from a scarlet to a white color, and it has commenced suppurating as though the thing were dead and coming out. The whole tumor is very much flattened, the discharge is different and not near so offensive. The greatest improvement is in her complexion. From a *tallowy*, puffy-looking, and somewhat bluish skin, she is regaining her old natural look, the skin shrinking, becoming wrinkled and clear.

"I am so happy in the prospect of a cure that I feel like a new man, as though a ton of lead had been lifted from my heart. Is it not a little singular, it has not had any perceptible effect on her nervous system? Her digestion is good, and she begins to feel that she will get well.'

".On the fourteenth of the same month Mr. Matthews writes as follows:—

"'This is the seventeenth day since I commenced the use of cundurango; shall cease for a few days, and note carefully the effect. When I began the treatment, Mrs. Matthews' breast was almost as hard as a stone, about four inches in diameter, the cancer itself two inches in diameter, with raised edges, hard and scarlet-colored, bleeding profusely at the slightest touch, emitting an odor of the most sickening and disagreeable kind, discharging a brownish, cancerous, limpid fluid; the countenance bloated, tallowy-looking, with a bluish pallor of the whole face; the lips turned blue at the least exertion, so that I have been very much alarmed, fearing a rapid crisis and dissolution; at the same time the tumor itself enlarged with fearful rapidity, so much so that I could notice the growth from day to day.

"Now all is changed—the countenance has resumed its old, familiar look; she moves about with great sprightliness, the blue of the lips no longer indicating fatigue or effort. The glandular swelling under the chin is gone; strength increasing; the tumor itself much flattened and decreased in protuberance; the color changed to a white, maturating sore; the limpid cancerous discharge ceased, and in its place a healthy discharge of white matter much less offensive; the hardened glands are soft to the touch, the whole symptoms indicating most plainly to me that the treatment has, so far, neutralized the poison of the blood, and that another short campaign with cundurango will insure a complete cure.

"On the 2d of July I visited Mrs. Matthews at South Bend, and was indeed astonished at the rapid change which had taken place. The tumor had become soft, the color natural, the secondary glandular deposits had all disappeared. The improved complexion, muscular firmness and elasticity of spirits all pointed to an early and complete recovery.

"Mrs. Handy, residing on M street, in this city, was the next subject of experiment with the cundurango. This was a highly-typical and fearfully-advanced case of cancer uteri. The grayish color, unequal, irregular elevations of the ulcer edges, the sympathetic disturbance of the bladder, the paroxysms of intense pain, together with the hot, dry, shrivelled, yellow surface, the wasted muscles, sunken eyes, the small, quick, wiry pulse, revealed one of those sad cases where all hope of remedy fails.

"The cundurango, in the form of decoction, was administered first to Mrs. Handy on the 31st day of last June. A regular record has been kept from day to day, describing the least change of symptoms, but I have not the space to introduce it here. Suffice it that even in this extreme case the beneficial effects of this wonderful remedial agent have been most apparent. The pain has steadily declined, the diseased parts are less tumefied and sensitive, and the discharge is very slightly offensive. The cachectic appearance of this patient has much improved, and she expresses herself as feeling altogether better.

"A lady of the family of Hon. Mr. Gorham, Secretary of the United States Senate, has had mammary cancer of several months' duration, and her condition was pronounced hopeless by leading Northern surgeons. I was called to see her on the first of June of this year, and found cancer of the breast, with secondary deposits in the shoulder and humeral portion of the left arm, attended by extreme rigidity of the neck, and almost complete immobility of the affected limb.

"A careful daily record has been preserved of this case, also, by which the most decided improvement is indicated. The mammary tumor has grown softer, and the line of skin-attachment bisecting the nipple is much less marked. The head, before stiff, is now perfectly free and movable, while the natural mobility of the disabled arm is restored, and the tissues, before hard, are now soft and natural. The general condition progresses favorably *pari passu* with the local improvement."

Mr. Vice-President COLFAX himself, in a letter to a friend, speaks of the action of the cundurango in the case of his mother, Mrs. Matthews, as follows:—

"I am glad to be able to tell you that mother is really on the high road apparently to a perfect cure, although she has only taken about quarter doses of the cundurango, in consequence of its scarcity. When we left Washington, in April, her case was absolutely hopeless; cancer growing fearfully and angrily. * * * Now the tumor is three-fourths gone, and apparently diminishing; pain almost gone and every symptom favorable. Since the first fortnight she has had only quarter doses, and now she has none. She is more like herself than she has been for years. How it cures, or how it affects the cancer, I cannot imagine. I know how incredulous many doctors are about it, and I would be too if I had not seen its results. It seems to depurate from the blood whatever it is that causes cancer, and I don't know what that is any more than I know why Peruvian bark cures ague. You can tell your friends however, when they obtain it, they will notice on the fourth day the improvement, and by the ninth day they will see themselves that the cancer is going away—that is, if it acts with them as with the cases I have seen. I am longing for its arrival, and glad that Dr. Bliss so promptly sent his partner to that distant region for it. I have the most piteous appeals for it from friends, offering hundreds for it if it will only stop the growth of this terrible disease, but I have not an iota, and I guess all in the United States is now used up."

Of course the medical profession will fully test the merits of this new remedy, and it will take its proper place in the armamentaria of the physician, be it as a specific or as an adjunct.

INDEX.

CANCER:—(*General Subjects.*)

Cachexia,	173
Cauliflower,	150
Cells,	43, 110
Formation of,	94
Classification of,	107
Constitutional nature of,	21
Bleeding,	15
Defined,	94
Degeneration of,	111
Diagnosis of tumors of breast,	171
Epithelial,	111, 114
Experience of the Author,	109
Fibroid,	111
Formation of,	111, 116
Fungoid,	111, 116
Geological distribution of,	165
Glanders, and,	51, 53
Hematoid,	112
Irritating agents, producing,	13
Lepoid,	166
Lupus,	155, 168
Microscopic characteristics,	32, 33, 43
Examinations of suspected,	97
Observations on, by the Author,	137
Pathological, changes in,	49
Patients, average duration of life in,	162
Physiological history of,	39
Pathology of,	41
Record of cases,	119
Scrofula and,	51, 59
Skin, how affected by,	172
Tubercle and,	51, 53, 61, 65, 85

CANCER:—(*Special Subjects.*)

Of breast,	77
Esophagus,	170
Internal organs,	170
Neck,	36
Uterus,	17, 72, 78, 163, 165

(See also Record of cases, pp. 119–135.)

CANCER:—(*Treatment by*)

 Caustics, - - - - - - - 69
 Cautery in uterine, - - - - - 73
 Cold, - - - - - - - 31, 67
 Compression, - - - - - 23, 25
 Excision, - - - - - - 31, 67
 Escharotics and styptics, - - - - 79
 Refrigeration, - - - - - 31, 67

CANCER:—(*Remedies.*)

 Preparation and composition of, - - 118
 Acid, nitric, - - - - - - 68, 115
  Sulphuric and saffron, - - - 70
  Acetic, - - - - - 80, 174, 178
  Citric and carbolic, - - - - 178
 Arsenic, - - - - - 69, 76, 78
  Iodide of, - - - - - - 20
 Anodyne applications in uterine, - - 20
 Bromine, - - . - - - 83, 152
  Chloride of and basilicon in Lupus, - 155
 Coneine, - - - - - - 11
 Colloid, styptic, - - - - 77, 83
 Donovan's solution, - - - - 75
 Gallium aperine, - - - - - 71
 Gold, chloride of, - - - - - 155
 Hemlock, - - - - 11, 66, 157
 Hip-bath in uterine, - - - - - 20
 Iodine, - - - - - - - 71
 Iron, perchloride of, in uterine, - - 72
 Opium as a palliative in, - - - - 77
 Paste, No. 1, No. 3, - - - - - 119
  Vienna, - - - - - - 154
 Podophyllum, extract of, - - - - 119
 Potash, chlorate of, - - - - - 70
  Iodide of, - - - - - - 71
 Richardson's styptic, - - - - - 77
  Recipe for, - - - - - - 174
 West's, Dr., prescription, - - - - 71
 Zinc, chloride of, - - - - 76, 119

AUTHORS QUOTED.

Arnott, Dr.,	23, 25, 67
Atlee, W. L.,	78
Barclay, J.,	178
Belajeff, Dr.,	96
Bennett, J. H.,	31, 33, 174
Bouley,	63
Broadbent, Dr.,	80
Carmichael, Mr.,	74
Collis, Maurice H.,	140
Cook, Weeden,	79
Devay, M.,	69
Ellis, Mr.,	73
Erichsen, Dr.,	169
Fanchou, Dr.,	68
Geuner,	60
Gross, S. D.,	109, 166
Haviland, Dr.,	105
Henry, A.,	147
Johns, Robert,	72
Landolfi, Dr.,	154
Lebert,	62
Lefranc,	12
Manec, M.,	69
Madden, Dr.,	79
Montgomery, Dr.,	17, 95
Morton, James,	174
Osborn, Dr.,	12, 66
Paget, J.,	162
Phillippe, Dr.,	63
Revallie, Dr.,	68
Routh, Dr.,	83
Salisbury, Dr.,	105
Solly, Samuel,	154
Thomas, T. Gaillard,	163
Villemin,	57
Virchow,	57
Walshe, W. H.,	21, 23
Young, Samuel,	23

www.ingramcontent.com/pod-product-compliance
Lightning Source LLC
Chambersburg PA
CBHW020846160426
43192CB00007B/799